INDUSTRIAL MECHANICS

Second Edition

Workbook

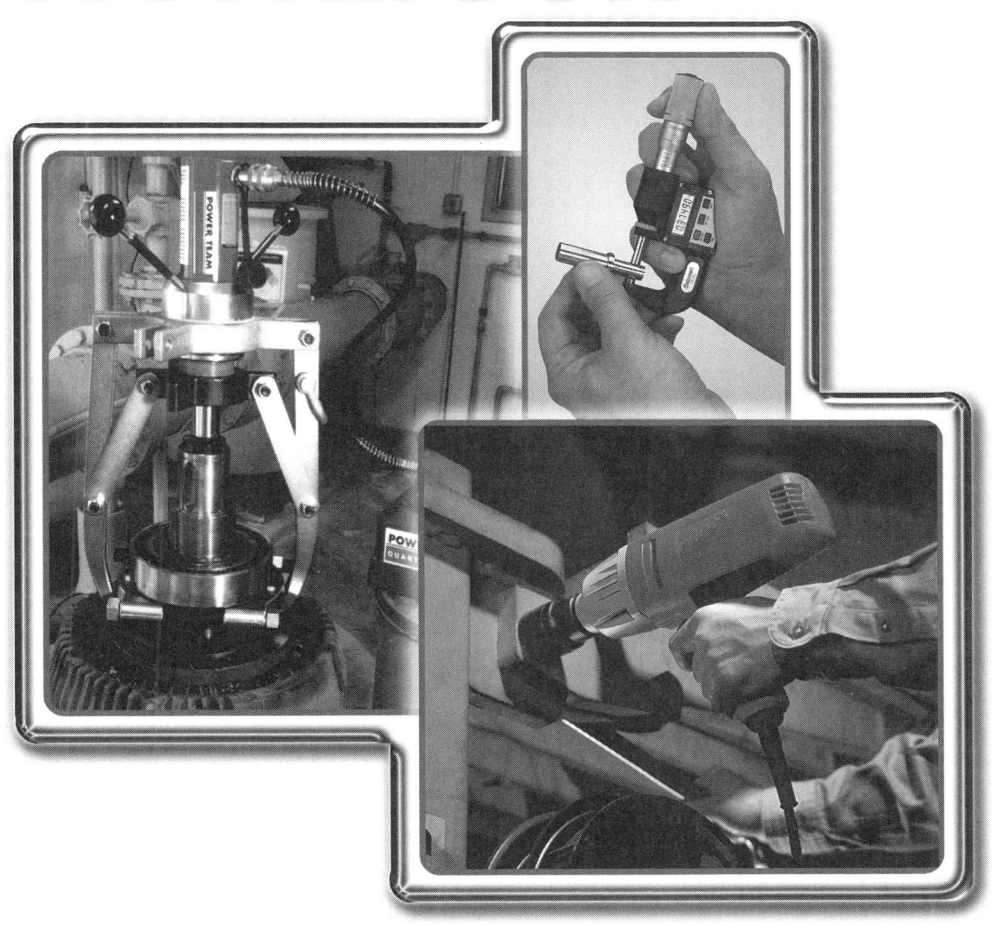

AMERICAN TECHNICAL PUBLISHERS, INC.
HOMEWOOD, ILLINOIS 60430-4600

ATP Staff

Industrial Mechanics contains procedures commonly practiced in industry and the trade. Specific procedures vary with each task and must be performed by a qualified person. For maximum safety, always refer to specific manufacturer recommendations, insurance regulations, specific job site and plant procedures, applicable federal, state, and local regulations, and any authority having jurisdiction. The material contained is intended to be an educational resource for the user. American Technical Publishers, Inc. assumes no responsibility or liability in connection with this material or its use by any individual or organization.

National Electric Code® (NEC®), and NEC are registered trademarks of the National Fire Protection Association, Inc., Quincy, MA 02269. MasterFormat is a registered trademark of The Construction Specifications Institute, Inc., Alexandria, VA 22314, Vise-Grip is a registered trademark of American Tool Companies, Inc., Vernon Hills, IL 60061, Teflon is a registered trademark of E. I. Du Pont de Nemours and Company, Wilmington, DE 19898, and QUAD-RING is a registered trademark of Quadion, Minneapolis, MN 55441

© 2008 by American Technical Publishers, Inc.
All rights reserved

2 3 4 5 6 7 8 9 – 08 – 9 8 7 6 5 4 3 2 1

Printed in the United States of America

    ISBN 978-0-8269-6999-8

 This book is printed on 30% recycled paper.

CONTENTS

1 Precision Measurement — 1

2 Printreading — 7

3 Tools — 13

4 Calculations — 21

5 Rigging — 33

6 Lifting — 41

7 Ladders and Scaffolds — 47

| **8** | **Hydraulic Principles** | **53** |

| **9** | **Practical Hydraulics** | **63** |

| **10** | **Pneumatic Principles** | **75** |

| **11** | **Practical Pneumatic** | **81** |

| **12** | **Lubrication** | **87** |

| **13** | **Bearings** | **97** |

| **14** | **Flexible Belt Drives** | **103** |

| **15** | **Mechanical Drives** | **109** |

16 Vibration 115

17 Alignment 121

18 Electricity 127

Final Exam 133

Appendix 143

INTRODUCTION

Industrial Mechanics Workbook is designed to reinforce the concepts in, and provide application activities for, the material presented in *Industrial Mechanics*. When studying the textbook, focus on italicized terms, illustrations, and examples. These key elements comprise a major portion of the workbook.

Review Questions

The workbook contains 23 sections of Review Questions. Each section of Review Questions is a series of multiple choice, true/false, completion, and matching questions based on the text and illustrations in the corresponding chapter of the textbook. Study the assigned chapter of the textbook thoroughly before completing the Review Questions.

Problems

The workbook contains 145 Problems developed from the 18 chapters of the textbook. Problems provide opportunities to apply the concepts and theory in the textbook to practical mechanical and troubleshooting problems. Most Problem solutions require the application of basic math skills.

Final Exam

The Final Exam is developed from a selection of Review Questions from each of the chapters. The Final Exam is designed to test basic knowledge of mechanics, industrial systems, and troubleshooting.

Appendix

The comprehensive Appendix contains many useful tables, charts, and other supplemental reference material that is helpful for developing solutions for each of the Problems and when students begin working in industrial facilities and continue learning with on-the-job training. See page 143 for a complete listing of material in the Appendix.

Related Information

Information presented in *Industrial Mechanics* and *Industrial Mechanics Workbook* addresses common mechanical topics for several types of industrial systems. Answers to the workbook Review Questions and Problems are provided in the *Industrial Mechanics Answer Key*. *Industrial Mechanics* is one of many high-quality training products available from American Technical Publishers, Inc. To obtain information about related training products, visit the American Tech web site at www.go2atp.com.

The Publisher

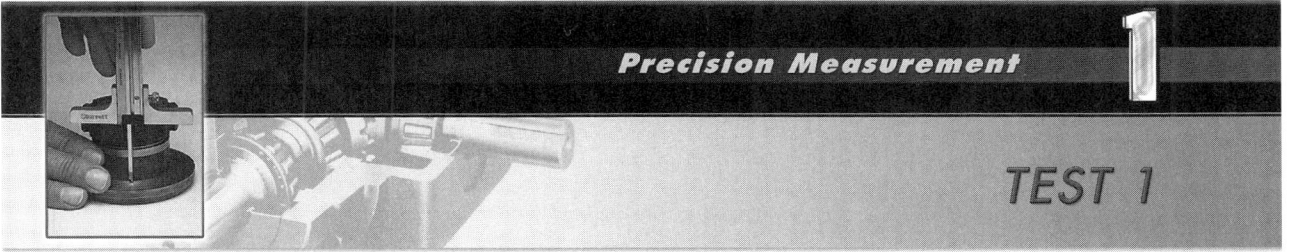

Precision Measurement

TEST 1

Name _____ Date _____

Industrial Mechanics

_____ 1. A(n) ___ is a tool designed to measure printed angles.

T   F   2. Electronic calipers are more precise and easier to read than vernier or dial calipers but require a DC power source (battery) for operation.

T   F   3. An inside micrometer is a micrometer used for measuring outside diameters and thicknesses of parts.

T   F   4. A vernier scale is a scale that has ten spaces and eleven lines.

_____ 5. A(n) ___ is a straight line or lines intersecting a point (vertex) of an angle.

_____ 6. A ___ is a permissible deviation from a given value or dimension.
   A. variation
   B. tolerance
   C. rule
   D. none of the above

T   F   7. Angles are measured in inches or centimeters.

_____ 8. Depth micrometers are used to measure dimensions of workpieces that have critical inside dimensions, such as ___.
   A. splines
   B. filter housings
   C. sleeve bearings
   D. all of the above

_____ 9. ___ is the verification of graduations and incremental values of a precision measuring instrument for accuracy and adjustments.
   A. Tolerance
   B. Caliper adjustment
   C. Calibration
   D. Maintenance

_____ 10. A(n) ___ is a measuring tool divided into even increments.

1

T F 11. A machinist's steel protractor is a tool used to measure or mark angle measurements on flexible, pliable workpieces.

_____ 12. A ___ is a tool for measuring and laying out angles.
A. caliper
B. rule
C. protractor
D. micrometer

_____ 13. Micrometers are used for verification of component dimensions such as ___.
A. thickness
B. diameter
C. depth
D. all of the above

_____ 14. A(n) ___ is the fixed measuring surface of a micrometer.
A. anvil
B. thimble
C. stop
D. spindle

_____ 15. A(n) ___ micrometer is a micrometer with a digital electronic indicating gauge.

T F 16. Precision measuring instrument handling requires proper cleaning and lubrication of the instrument before and after each use.

T F 17. Precision test instruments will not give faulty readings because of any expansion or contraction of the test instrument caused by temperature extremes.

_____ 18. ___ is a method of using measuring instruments to acquire accurate measurements.

_____ 19. ___ are used to make precision inside and outside measurements on components, such as measurements of inside diameters (ID), outside diameters (OD), and length.

_____ 20. A(n) ___ angle is an angle exceeding 90°, and less than 180°.

T F 21. An angle is a geometric figure formed by two lines extending from the same point.

T F 22. An inside micrometer is a micrometer used to measure linear dimensions between two inside points or parallel surfaces.

_____ 23. A(n) ___ is a finely ground, precisely sized object that is used as a basis of dimensional comparison.

_____ 24. A ___ is usually one tool with a combination of different measuring devices attached to a steel rule and is sometimes referred to as a combination square set.
A. caliper
B. rule
C. protractor
D. reversible protractor

25. Conventional protractors are ___ in shape and have an outer scale, inner scale, zero edge, and center mark.
    A. circular
    B. semicircular
    C. rectangular
    D. oblong

26. Depth micrometer measuring rod extensions typically vary by exactly ___" and are calibrated by the set manufacturer.
    A. ¼
    B. ½
    C. 1
    D. 2

27. ___ are designed with standard micrometer heads (thimble and barrel) attached to a flat "tee" base.
    A. Inside micrometers
    B. Outside micrometers
    C. Calipers
    D. none of the above

28. Micrometers with a vernier scale are capable of taking measurements to the nearest ___".
    A. 0.01
    B. 0.001
    C. 0.0001
    D. 0.00001

29. Electronic calipers can indicate readings in thousandths of an inch or ___ of a millimeter.
    A. tenths
    B. fiftieths
    C. hundredths
    D. thousandths

30. A micrometer used to measure grooves or keyways has a ___-shaped anvil and spindle.
    A. ball
    B. disc
    C. point
    D. cup

4 INDUSTRIAL MECHANICS WORKBOOK

Dial Calipers

_____ 1. Adjustment screw

_____ 2. Calibration screw

_____ 3. Depth measurement blade

_____ 4. Dial scale

_____ 5. Inside jaws

_____ 6. Locking screw

_____ 7. Main beam (scale)

_____ 8. Outside jaws

_____ 9. Sliding mechanism

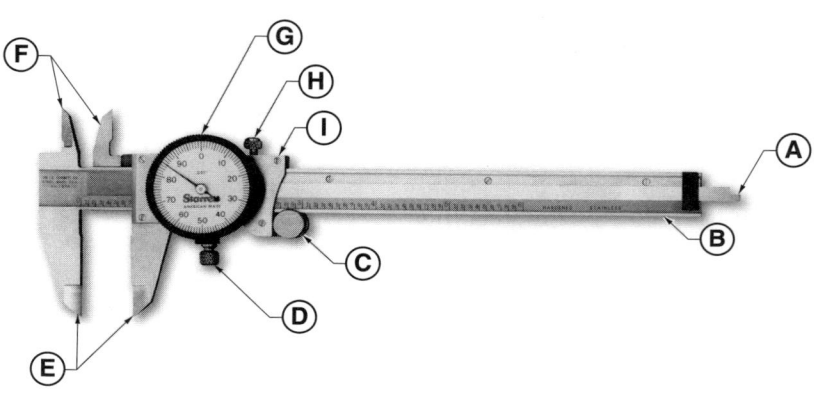

Outside Micrometers

_____ 1. Anvil

_____ 2. Barrel

_____ 3. Barrel scale

_____ 4. Frame

_____ 5. Spindle

_____ 6. Spindle movement

_____ 7. Stop

_____ 8. Thimble

_____ 9. Thimble rotation

_____ 10. Thimble scale

_____ 11. Vernier scale

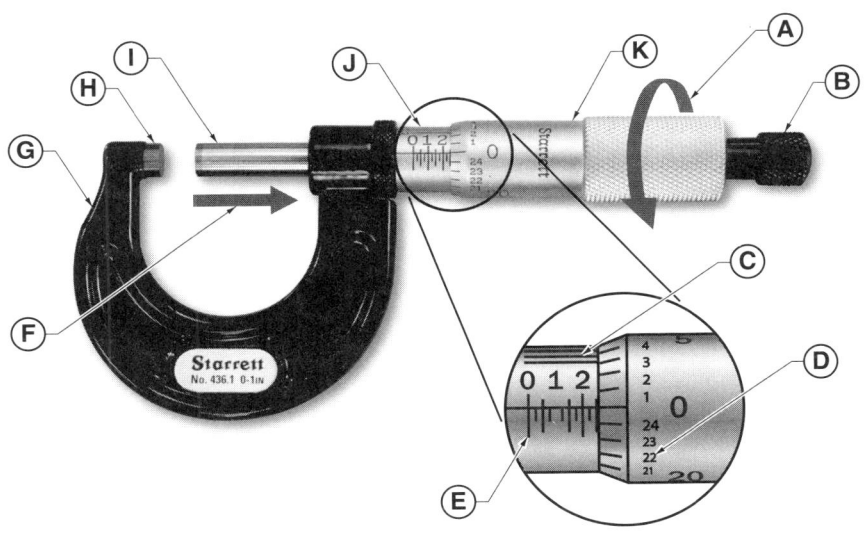

Depth Micrometers

_____ 1. Base

_____ 2. Barrel

_____ 3. Measured shoulder

_____ 4. Rod

_____ 5. Stop

_____ 6. Thimble

_____ 7. Zero point

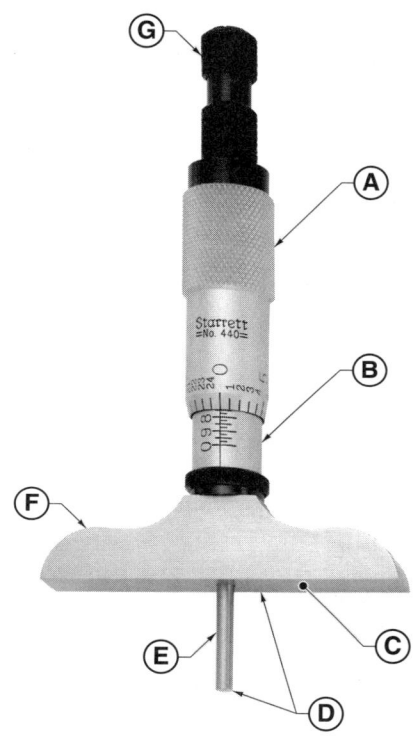

Printreading 2

TEST 1

Name _____ Date _____

Industrial Mechanics

_____ 1. A(n) ___ is a drawing that shows the property lines of a building lot, elevation, compass directions, lengths of property lines, and locations of structures.

_____ 2. A(n) ___ is a line that is used with a written dimension to indicate size or location.

_____ 3. A(n) ___ is a system of drawing representation in which drawing elements are proportional to actual elements.

T   F   4. Foundation plans are used to determine the building materials that are used to build the main support structure.

T   F   5. Common architect's scales are rectangular in cross section and have six edges.

_____ 6. A floor plan is a plan view looking down at a building from approximately ___′ above the floor.
  A. 5
  B. 10
  C. 50
  D. 100

_____ 7. A(n) ___ drawing is a type of drawing that is used to indicate how to do work using the simplest and/or safest method.
  A. assembly
  B. sectional
  C. instructional
  D. detail

_____ 8. ___ plan designs include information about the landscaping that a specific piece of property can have.
  A. Floor
  B. Foundation
  C. Structural
  D. Utility

_____ 9. All prints are composed of ___ to show the shape of the drawn object.

7

8 INDUSTRIAL MECHANICS WORKBOOK

_____ 10. A(n) ___ is a block that identifies the changes that have been marked on the drawing since its initial approval.

_____ 11. A ___ is a reproduction of original drawings created by an architect or engineer.
  A. print
  B. plan
  C. legend sheet
  D. note

T   F   12. The two types of notes are general (construction) notes and sheet notes.

_____ 13. A(n) ___ is a type of drawing where all faces of an object are projected onto flat planes that generally are at 90° angles to one another.

_____ 14. ___ plans are used to provide excavating, construction, drainage, water-proofing, and other design information for building the foundation.

_____ 15. A(n) ___ is an assembly of lines, dimensions, and notes used to convey general or specific information as required by the application and use.

_____ 16. ___ drawings are often used in installation and operational manuals to show where to connect external wires and position indicating lamps, switches, and displays.
  A. Application
  B. Orthographic
  C. Location
  D. none of the above

_____ 17. Title block information typically includes ___.
  A. subject or sheet contents
  B. project title and location
  C. print division and print number
  D. all of the above

T   F   18. A break line is used to show the alternate positions of a repeating detail.

T   F   19. A head-on view is a view when looking directly at an object from the same height as the object.

T   F   20. A sectional drawing is a type of drawing that indicates the internal features of an object.

_____ 21. A(n) ___ is a type of drawing that indicates the location of utilities such as electrical, water, sewage, gas, and communication cables.

_____ 22. A(n) ___ is additional information that is included with a set of prints.

_____ 23. A(n) ___ shows how the individual parts of an object work together.

_____ 24. Plans are two-dimensional drawings designed to indicate the ___ of objects.

_____ 25. A ___ drawing is a three-dimensional drawing that resembles a picture.
  A. head-on
  B. location
  C. detail
  D. pictorial

_____ 26. ___ are used to locate the centers of windows, doors, and electrical enclosures, and to indicate that an object is round or cylindrical in shape.
  A. Dimension lines
  B. Title blocks
  C. Centerlines
  D. all of the above

_____ 27. General-purpose section lines are typically drawn on a 45° angle and ___″ apart.
  A. 1/32
  B. 1/16
  C. 1/10
  D. 1/8

_____ 28. Underground utilities are commonly indicated with a ___ line or with a solid line that is broken for placement of a letter indicting the type of utility.
  A. dashed
  B. dotted
  C. dotted and dashed
  D. solid

_____ 29. Small sets of prints are used for reference purposes and are usually either 8″ × 11″ or ___″ × ___″.
  A. 8, 10
  B. 8, 12
  C. 11, 17
  D. 11, 24

_____ 30. A standalone symbol "×" on a print is an abbreviation for ___.
  A. centerline
  B. diameter
  C. radius
  D. repetitive feature

Section Line Types

_____ 1. Electrical windings electromagnets, resistance, etc.

_____ 2. Marble, slate, glass, porcelain, ceramics

_____ 3. Cast iron, malleable iron, and general use for all materials

_____ 4. Wood

_____ 5. Magnesium, aluminum, and aluminum alloys

_____ 6. Steel

_____ 7. Rubber, plastic, and electrical insulation

_____ 8. Sound insulation

_____ 9. Titanium and refractory metal

_____ 10. Rock

_____ 11. Cork, felt, fabric, leather, and fiber

_____ 12. Earth

_____ 13. Sand

_____ 14. Concrete

_____ 15. Thermal insulation

_____ 16. Water and other liquids

_____ 17. Bronze, brass, copper

_____ 18. White metal, zinc, lead, babbitt, and alloys

Line Types

1. Draw an object line.

2. What is an object line used for in prints?

3. Draw a phantom line.

4. What is a phantom line used for in prints?

5. Draw a hidden line.

6. What is a hidden line used for in prints?

Tools 3 — TEST 1

Name _____ Date _____

Industrial Mechanics

_____ 1. A(n) ___ is a steel hand tool with one end formed to a conical point of approximately 90°.

_____ 2. A(n) ___ is used to smooth areas enclosed by an acute angle.

T   F   3. There are three main types of screwdrivers: flat head, Phillips head, and Allen.

T   F   4. Chisel types used for industrial and mechanical applications include flat cold, cape, round nose, and diamond point.

T   F   5. One of the guidelines for using hand tools for industrial applications is replacing damaged or worn tools as required.

_____ 6. A common wrench used for industrial and mechanical applications is the ___ wrench.
       A. socket
       B. adjustable
       C. pipe
       D. all of the above

_____ 7. ___ taps are used after a taper tap has been used to start a true and straight thread.
       A. Bottom
       B. Plug
       C. Die
       D. Wrench

_____ 8. The main parts of a handsaw are the blade, teeth, back, and ___.
       A. chisel
       B. head
       C. handle
       D. none of the above

_____ 9. A(n) ___ is a striking or splitting tool with a hardened head fastened perpendicular to a handle.

_____ 10. Most ___ strip standard wire from AWG size 22 to AWG size 10 and solid wire from AWG size 18 to AWG size 8.

_____ 11. Industrial and mechanical technicians use ___ for various gripping, turning, cutting, positioning, and bending operations.

_____ 12. Some ___ resemble reverse-threaded screws, while others resemble square tapered rods with chiseled edges.
   A. screw extractors
   B. vises
   C. tape rules
   D. mechanical pullers

_____ 13. A ___ is a hand tool with a tip designed to fit into a screw head for fastening operations.
   A. handsaw
   B. vise
   C. screwdriver
   D. punch

T   F   14. Vises usually consist of a screw, lever, or cam mechanism that closes and holds two or more jaws around a workpiece.

T   F   15. High-quality handsaw blades should not be tapered from the blade top to the blade bottom.

_____ 16. End-cutting pliers are used for cutting ___ close to the workpiece.
   A. wire
   B. nails
   C. rivets
   D. all of the above

_____ 17. File parts include the point, edge, face, heel, and ___.
   A. handle
   B. head
   C. tang
   D. blade

_____ 18. A(n) ___ is a tool used to remove fitted machine parts.

_____ 19. ___ due to improper use of taps when cutting metal can cause the cutting edges of taps to chip, break, or shatter.

_____ 20. A(n) ___ is a cut that is made against the direction of the wood grain and is made with full, even strokes at about a 45° angle.

T   F   21. External pullers are used to remove objects such as bearings or bushings from a bore.

T   F   22. Dies have a side with a 90° chamfer to prevent die-tooth breakage and to allow for a gentle cutting start.

T   F   23. Some types of power drills, such as hammer drills, can drill at speeds up to 3000 rpm and simultaneously hammer into the material at up to 50,000 blows per minute.

_____ 24. ___ are used with a tap wrench to "pull" the tap into a workpiece.
    A. Taps
    B. Pliers
    C. Files
    D. Mechanical pullers

_____ 25. ___ dies are used for tough cutting and are designed with thicker cross sections than round dies to permit cleaning and rethreading of threads.
    A. Square
    B. Triangular
    C. Hexagonal
    D. Octagonal

_____ 26. A ___-handle hacksaw is usually preferred for fine work.
    A. pistol
    B. straight
    C. wooden
    D. metal

T   F   27. A circular saw is a multipurpose cutting tool in which the blade reciprocates to create the cutting action.

_____ 28. Single-cut files have a single set of teeth and are used to make cuts at an angle between ___° and ___°.
    A. 45, 90
    B. 65, 85
    C. 75, 85
    D. 75, 90

_____ 29. An 18-point saw blade has larger teeth than a ___-point saw blade.
    A. 8
    B. 10
    C. 16
    D. 20

_____ 30. Reciprocating saws typically operate at ___ to ___ strokes per minute (no load).
    A. 1200, 1800
    B. 1500, 2500
    C. 1700, 2800
    D. 1800, 2900

Circular Saw Blades

_____ 1. Abrasive

_____ 2. Carbide-tipped finish and trim

_____ 3. Carbide-tipped framing/rip

_____ 4. Carbide-tipped plywood/veneer

_____ 5. Chisel-tooth combination

_____ 6. Combination

_____ 7. Metal-cutting

_____ 8. Plywood/paneling

_____ 9. Rip

Milwaukee Electric Tool Corp.

Punches

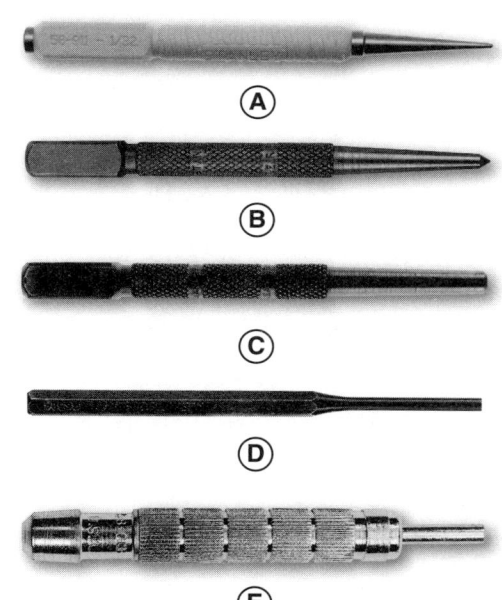

_____ 1. Center

_____ 2. Pin

_____ 3. Prick

_____ 4. Spring-loaded

_____ 5. Solid

The Stanley Works

Pliers

_____ 1. Diagonal-cutting _____ 5. Long nose

_____ 2. End-cutting _____ 6. Self-adjusting

_____ 3. Lineman's _____ 7. Slip-joint

_____ 4. Locking _____ 8. Tongue-and-groove

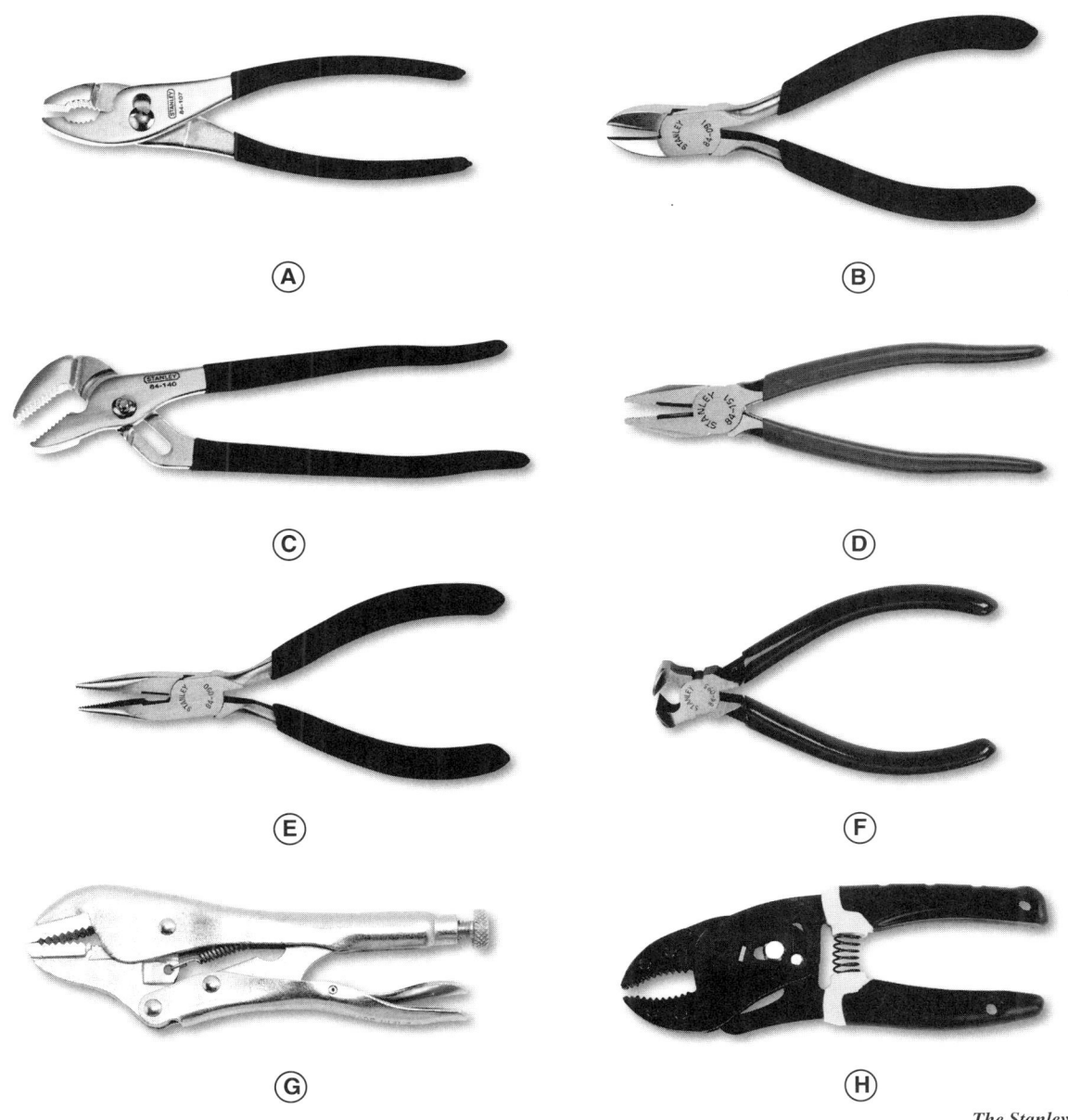

The Stanley Works

File Shapes

_____ 1. Flat

_____ 2. Half-round

_____ 3. Knife

_____ 4. Mill

_____ 5. Round

_____ 6. Square

_____ 7. Three-square

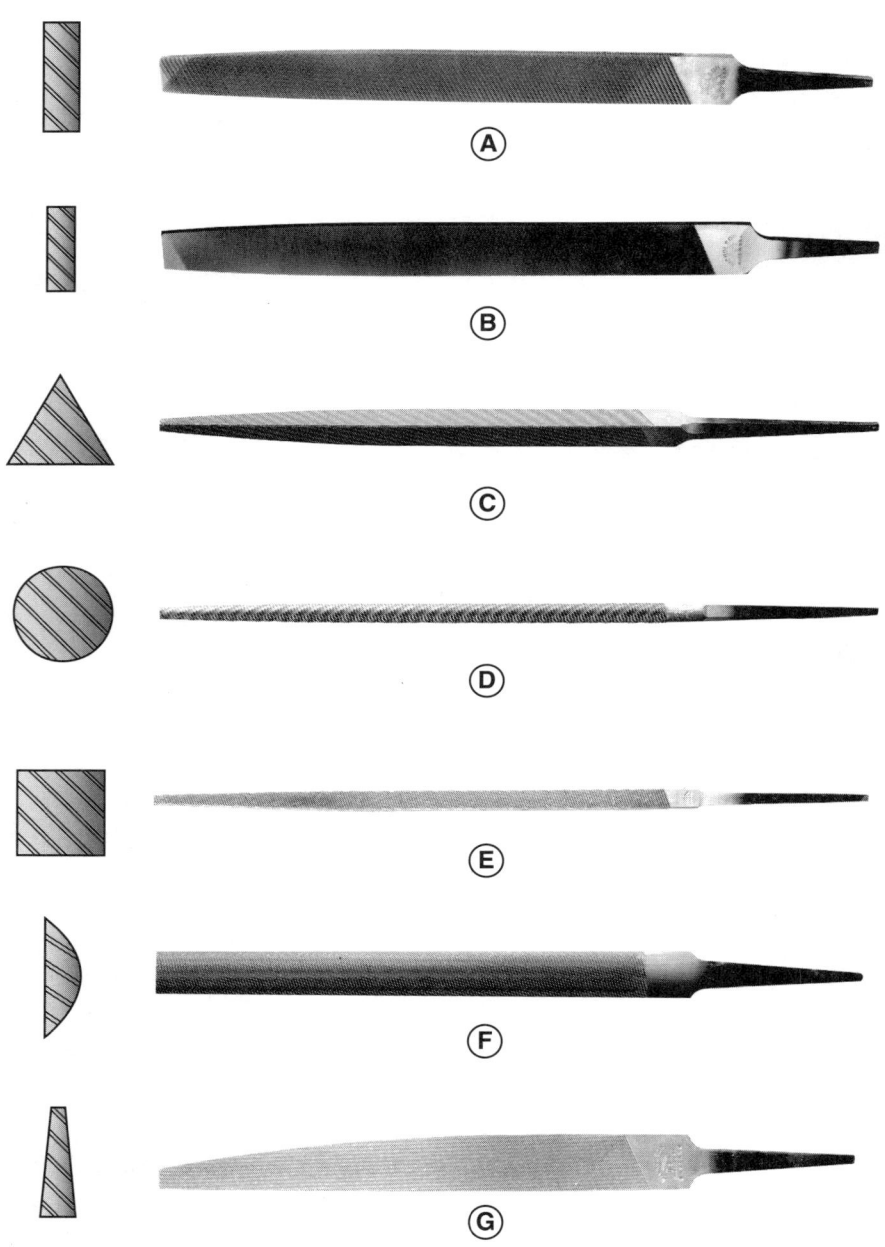

Cooper Industries, Inc.

Tool Usage

1. Describe applications where a flat file is typically used.

2. Describe the procedure for removing a broken bolt or screw with a screw extractor.

3. The tap-drill size for a ½″-22 threaded hole is ___″.

4. Describe the procedure for cutting external threads by hand with a single-piece die.

5. Describe the procedure for cutting material by hand with a handsaw.

Calculations 4

TEST 1

Industrial Mechanics

_____ 1. A(n) ___ is a means of showing that two numbers or two groups of numbers are equal to the same amount.

_____ 2. A(n) ___ figure is a flat figure with no depth.

_____ 3. A(n) ___ is the intersection of two lines or sides.

T F 4. All circles contain 360°.

T F 5. The sum of the three angles of a triangle is always 90°.

_____ 6. A quadrilateral always ___.
   A. has four sides
   B. contains 360°
   C. both A and B
   D. none of the above

_____ 7. The ___ of a prism is the perpendicular distance between the two bases.

T F 8. A right cylinder is a cylinder with the axis perpendicular to the base.

_____ 9. A(n) ___ is a solid generated by a circle revolving about one of its axes.

_____ 10. A(n) ___ is a mathematical equation that contains a fact, rule, or principle.

_____ 11. ___ is the number of unit squares equal to the surface of an object.

_____ 12. The ___ of a triangle is the side upon which the triangle stands.

_____ 13. A(n) ___ is a quadrilateral with all sides equal and four 90° angles.

_____ 14. A ___ is a quadrilateral with opposite sides equal and four 90° angles.
   A. square
   B. rectangle
   C. rhombus
   D. rhomboid

_____ **15.** A ___ is a quadrilateral with all sides equal and no 90° angles.
   A. square
   B. rectangle
   C. rhombus
   D. rhomboid

_____ **16.** A(n) ___ is a quadrilateral with opposite sides equal and no 90° angles.

    T   F   **17.** A formula can be changed to solve for any unknown value if the other values are known.

    T   F   **18.** A square foot contains 12 sq in.

    T   F   **19.** Polyhedra are solids bound by plane surfaces.

_____ **20.** A(n) ___ is a regular solid of eight triangles.
   A. hexahedron
   B. octahedron
   C. tetrahedron
   D. dodecahedron

_____ **21.** A(n) ___ is a regular solid of twelve pentagons.
   A. hexahedron
   B. octahedron
   C. tetrahedron
   D. dodecahedron

_____ **22.** A(n) ___ is a regular solid of six squares.
   A. hexahedron
   B. octahedron
   C. tetrahedron
   D. dodecahedron

_____ **23.** A(n) ___ is a regular solid of four triangles.

_____ **24.** The prefix kilo (k) has a prefix equivalent of ___.

_____ **25.** A(n) ___ of a pyramid or cone is the remaining portion of a pyramid or cone with a cutting plane passed parallel to the base.

_____ **26.** ___ angles are two angles formed by three lines in which the sum of the two angles equals 90°.
   A. Acute
   B. Complementary
   C. Obtuse
   D. Right

_____ **27.** ___ angles have the same vertex and one side in common.
   A. Adjacent
   B. Chord
   C. Concentric
   D. Tangent

_____ **28.** The ___ is the boundary of a circle.
   A. arc
   B. circumference
   C. diameter
   D. radius

_____ **29.** The circumference of a 47″ D circle is ___″.
   A. 73.83
   B. 147.65
   C. 173.50
   D. 295.31

_____ **30.** The circumference of a 47″ R circle is ___″.
   A. 147.65
   B. 173.50
   C. 295.31
   D. 1734.95

_____ **31.** The area of a 35″ D circle is ___ sq in.
   A. 54.98
   B. 240.53
   C. 962.12
   D. 3848.45

_____ **32.** The area of a triangle with a 12″ base and a 15″ height is ___ sq in.
   A. 13.5
   B. 27
   C. 90
   D. 180

_____ **33.** The length of a hypotenuse of a triangle having sides of 23 m and 31 m is ___ m.

_____ **34.** The area of a 116′ × 54′ warehouse is ___ sq ft.

_____ **35.** The volume of a sphere that is 1′ 2″ in diameter is ___ cu ft.

Lines

_____ 1. Line   A. Shortest distance between two points

_____ 2. Straight line   B. Line that is slanted

_____ 3. Horizontal line   C. Two or more lines that remain the same distance apart

_____ 4. Vertical line   D. Line that is perpendicular to the horizon

_____ 5. Inclined line   E. Line that is parallel to the horizon

_____ 6. Parallel lines   F. Boundary of a surface

Angles

_____ 1. Complementary

_____ 2. Supplementary

_____ 3. Right

_____ 4. Straight

_____ 5. Acute

_____ 6. Obtuse

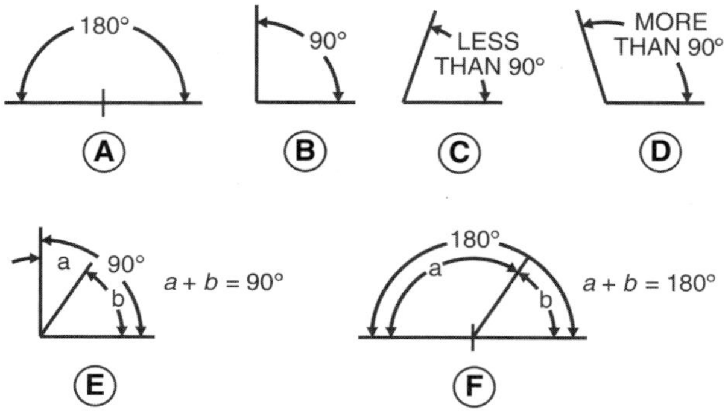

Circles

_____ 1. Centerpoint

_____ 2. Angle

_____ 3. Chord

_____ 4. Sector

_____ 5. Diameter

_____ 6. Semicircle

_____ 7. Segment

_____ 8. Radius

_____ 9. Quadrant

_____ 10. Arc

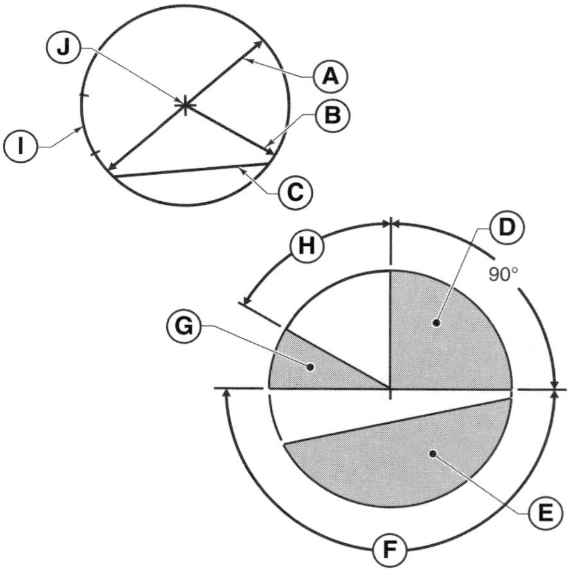

Polygons

_____ 1. Quadrilateral

_____ 2. Hexagon

_____ 3. Heptagon

_____ 4. Pentagon

_____ 5. Octagon

_____ 6. Triangle

Triangles

_____ 1. Isosceles

_____ 2. Equilateral

_____ 3. Right

_____ 4. Scalene (acute)

_____ 5. Scalene (obtuse)

Problems

_____ 1. The area of Circle A is ___ sq in.

_____ 2. The area of Circle B is ___ sq ft.

_____ 3. The circumference of Circle A is ___".

_____ 4. The circumference of Circle B is ___".

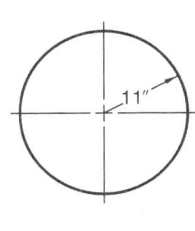

CIRCLE A

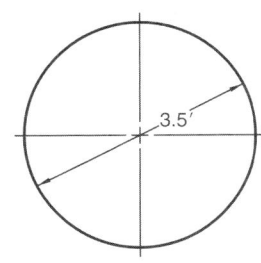

CIRCLE B

The piece of ⅛″ Rolled Steel is to be shear cut to produce 6″ × 8″ × 10″ triangular fins.

_____ **5.** The Rolled Steel has an area of ___ sq ft.

_____ **6.** The area of each triangular fin is ___ sq in.

_____ **7.** A total of ___ triangular fins can be produced from the Rolled Steel.

ROLLED STEEL

_____ **8.** The Storage Tank will hold ___ cu ft of water.

_____ **9.** The Holding Tank, Storage Tank, and pipe will hold ___ cu ft of water.

_____ **10.** A 2′ D solid concrete ball placed in the Storage Tank will displace ___ cu ft of water.

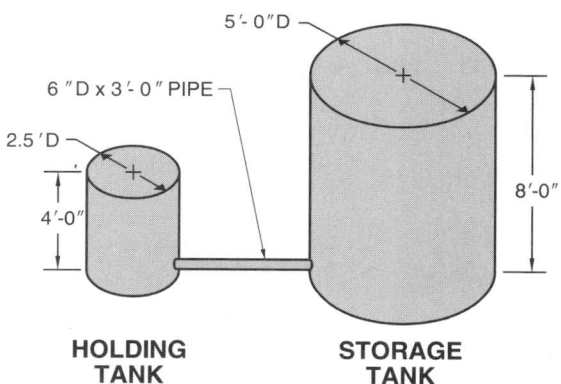

Calculations

TEST 2

Industrial Mechanics

_____ 1. A(n) ___ is the boundary of a surface.

_____ 2. A(n) ___ angle is two lines that intersect perpendicular to each other.

_____ 3. The ___ is the point of intersection of the sides of an angle.
   A. centerpoint
   B. radius
   C. vertex
   D. none of the above

_____ 4. A(n) ___ line is a line that is perpendicular to the horizon.
   A. horizontal
   B. vertical
   C. inclined
   D. none of the above

_____ 5. An acute angle is an angle that ___.
   A. contains less than 90°
   B. contains exactly 90°
   C. contains more than 90°
   D. may contain any number of degrees

T   F   6. There are 60′ in one degree.

T   F   7. A straight angle always contains 90°.

T   F   8. All lines may be drawn in any position unless they are horizontal or vertical.

_____ 9. A ___ is a portion of the circumference of a circle.
   A. vector
   B. chord
   C. segment
   D. none of the above

27

_____ **10.** Concentric circles are two or more circles with ___.
　　　　　　　　　　A. same diameters and same centerpoints
　　　　　　　　　　B. same diameters and different centerpoints
　　　　　　　　　　C. different diameters and same centerpoint
　　　　　　　　　　D. different diameters and different centerpoints

_____ **11.** Eccentric circles are two or more circles with ___.
　　　　　　　　　　A. same diameters and same centerpoints
　　　　　　　　　　B. same diameters and different centerpoints
　　　　　　　　　　C. different diameters and same centerpoint
　　　　　　　　　　D. different diameters and different centerpoints

_____ **12.** A(n) ___ is a three-sided polygon with three interior angles.

_____ **13.** The ___ is the side of a right triangle opposite the right angle.

_____ **14.** The angles of a triangle are named by ___ letters.

_____ **15.** The sides of a triangle are named by ___ letters.

　　T　　F　　**16.** Polygons are named according to their number of sides.

　　T　　F　　**17.** A right triangle has a 3-4-5 relationship.

_____ **18.** A(n) ___ is a solid with two bases that are parallel and identical polygons.

_____ **19.** A(n) ___ is a solid with a base that is a polygon and sides that are triangles.

_____ **20.** ___ is the three-dimensional size of an object measured in cubic units.

_____ **21.** A cubic inch measures ___ or its equivalent.
　　　　　　　　　　A. 1″
　　　　　　　　　　B. 1″ sq
　　　　　　　　　　C. 1″ × 1″
　　　　　　　　　　D. none of the above

_____ **22.** An obtuse triangle is a scalene triangle with ___.
　　　　　　　　　　A. one angle less than 90°
　　　　　　　　　　B. one angle of 90°
　　　　　　　　　　C. one angle greater than 90°
　　　　　　　　　　D. two angles of 90°

_____ **23.** A polygon is ___.
　　　　　　　　　　A. a many-sided plane figure
　　　　　　　　　　B. bound by straight lines
　　　　　　　　　　C. both A and B
　　　　　　　　　　D. none of the above

_____ 24. A trapezoid is a quadrilateral with ___ sides parallel.
　　　　　　　　　　　　A. no
　　　　　　　　　　　　B. two
　　　　　　　　　　　　C. opposite
　　　　　　　　　　　　D. all

_____ 25. The circumference of a sphere is equal to the circumference of a ___ circle.
　　　　　　　　　　　　A. great
　　　　　　　　　　　　B. small
　　　　　　　　　　　　C. either A or B
　　　　　　　　　　　　D. none of the above

T　　F　　26. In a formula, the sign of a number or letter is changed to the opposite sign when transposed.

T　　F　　27. Angles are measured in degrees, minutes, and seconds.

T　　F　　28. Supplementary angles are two angles formed by three lines in which the sum of the two angles equals 90°.

T　　F　　29. A chord is a line from circumference to circumference through the centerpoint of a circle.

T　　F　　30. A sector is a pie-shaped piece of a circle.

Pyramids

_____ 1. Right rectangular

_____ 2. Right triangular

_____ 3. Oblique pentagonal

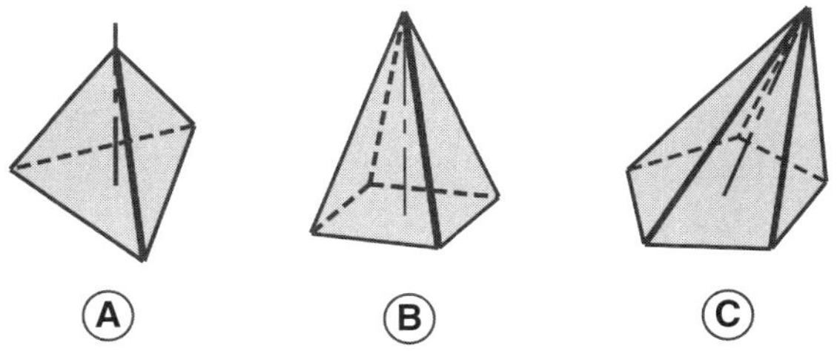

Quadrilaterals

_____ 1. Trapezium

_____ 2. Trapezoid

_____ 3. Square

_____ 4. Rectangle

_____ 5. Rhombus

_____ 6. Rhomboid

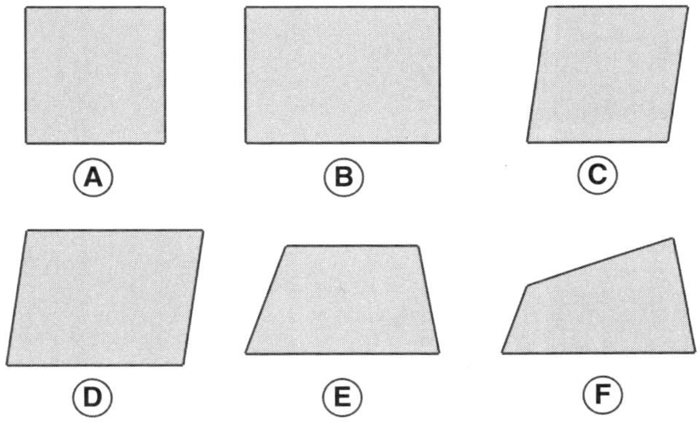

Regular Solids

_____ 1. Tetrahedron

_____ 2. Hexahedron

_____ 3. Octahedron

_____ 4. Dodecahedron

_____ 5. Icosahedron

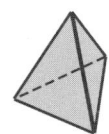

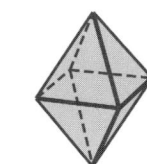
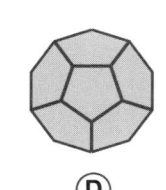
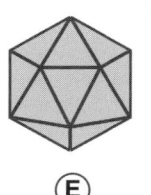

Other Regular Solids

_____ 1. Oblate ellipsoid

_____ 2. Prolate ellipsoid

_____ 3. Torus

_____ 4. Sphere

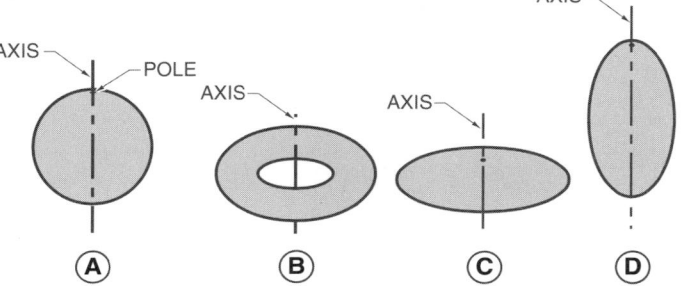

Conic Sections

_____ 1. Circle

_____ 2. Ellipse

_____ 3. Parabola

_____ 4. Hyperbola

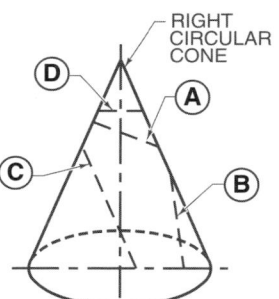

Problems

_____ 1. The area of Rectangle A is ___ sq in.

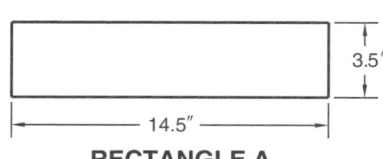

RECTANGLE A (14.5" × 3.5")

2. The Warehouse contains ___ sq ft.

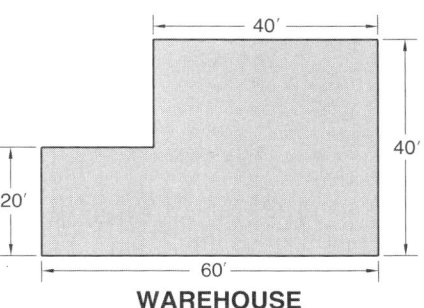
WAREHOUSE

3. The circumference of Circle A is ___′.
4. The area of Circle A is ___ sq ft.
5. The area of Triangle A is ___ sq in.
6. The length of Side c of Triangle A is ___″.
7. The area of Circle B is ___ mm².
8. The circumference of Circle B is ___ mm.
9. A circle has a 28″ diameter. The area of the circle is ___ sq in.
10. The Outer Zone contains ___ sq ft.

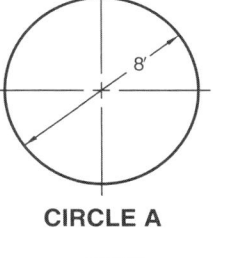
CIRCLE A

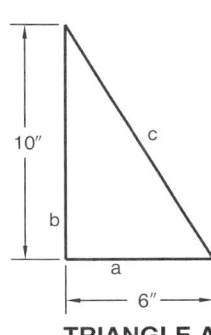
TRIANGLE A

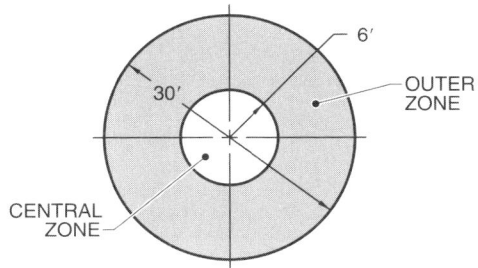

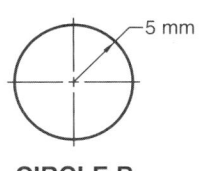
CIRCLE B

11. 2000 mg is equal to ___ dg.
    A. 0.02
    B. 0.2
    C. 2
    D. 200

12. 800 gal. is equal to ___ l.
    A. 2998
    B. 3010
    C. 3019
    D. 3028

_____ **13.** 500 mi is equal to ___ km.
   A. 310
   B. 312.5
   C. 800
   D. 804.5

_____ **14.** 10,000 lb is equal to ___ kg.
   A. 3730
   B. 3932
   C. 4533
   D. 4540

_____ **15.** 36 in. is equal to ___ cm.
   A. 91.34
   B. 91.44
   C. 91.54
   D. 91.64

_____ **16.** 1500 l is equal to ___ kl.
   A. 0.15
   B. 1.5
   C. 15
   D. 150

_____ **17.** Seven tons is equal to ___ lb.
   A. 14,000
   B. 28,000
   C. 35,000
   D. 42,000

_____ **18.** One square mile is equal to ___ A.
   A. 330
   B. 640
   C. 950
   D. 1260

_____ **19.** 937 dal is equal to ___ l.
   A. 0.937
   B. 9.37
   C. 93.7
   D. 9370

_____ **20.** One metric ton is equal to ___ g.
   A. 10,000
   B. 100,000
   C. 1,000,000
   D. 10,000,000

Rigging — TEST 1

Name _____ Date _____

Industrial Mechanics

_____ 1. ___ is securing equipment or machinery in preparation for lifting by means of rope, chain, or webbing.

_____ 2. A(n) ___ load is a load in which one-half of the load is a mirror image of the other half.

_____ 3. A lifting ___ is a thick metal loop (eyebolt) welded or screwed to a machine to allow balanced lifting.

T   F   4. The sling apex is the uppermost point where sling legs meet.

T   F   5. Rope is used for lifting because of its length and flexibility.

_____ 6. Rope ___ is the length of rope in which a strand makes a complete helical wrap around the core.

_____ 7. A(n) ___ is any kind of a class of sour substances with a pH value less than 7.

_____ 8. A(n) ___ is a curved piece of metal around which the rope is fitted to form a loop.

T   F   9. A socket is a rope attachment through which a rope end is terminated.

T   F   10. Because a rope is flexible, bending does not subject it to stress.

T   F   11. Fiber rope can be made from either natural or synthetic fibers.

_____ 12. A ___ is the interlacing of rope to form a permanent connection.
   A. hitch
   B. knot
   C. bight
   D. none of the above

_____ 13. ___ is a knitted or woven edge of a webbing formed to prevent raveling.
   A. Web ply
   B. Rebanding
   C. Selvedge
   D. Loop eye

33

_____ 14. ___ is the process in which metal is brought to a temperature below its critical temperature and allowed to cool slowly.

_____ 15. ___ strength is a metal's resistance to a force applied parallel to its contacted plane.

_____ 16. A steel ___ is a metallic material formulated from the fusing of two or more metals.

_____ 17. ___ wire is a wire rope constructed of strands consisting of more than one size wire staggered in layers.

_____ 18. A(n) ___ is the joining of two rope ends to form a permanent connection.

_____ 19. The web sling ___ is the distance between the extreme points of a web sling, including any fittings.

_____ 20. A(n) ___ is a U-shaped metal link with the ends drilled to receive a pin or bolt.

_____ 21. Regarding wire rope, a strength safety factor of ___ is used for steady or even loads.

T   F   22. Fiber core wire rope slings of all grades shall be permanently removed from service if they are exposed to temperatures exceeding 150°F.

T   F   23. A scaffold hitch is made from a clove hitch and a bowline knot.

T   F   24. The NACM specifies that the grade number or letter of a chain must appear at least once every 48 links.

T   F   25. Hitches work by the pressure of rope being pressed together.

Webbing

_____ 1. Tapered eye

_____ 2. Loop eye length

_____ 3. Eye width

_____ 4. Sling width

_____ 5. Web face

_____ 6. Selvedge

_____ 7. Splice

_____ 8. Warning core

_____ 9. Body

_____ 10. Length

Hooks

_____ 1. Swivel

_____ 2. Gated

_____ 3. Ungated

_____ 4. Eye

_____ 5. Clevis

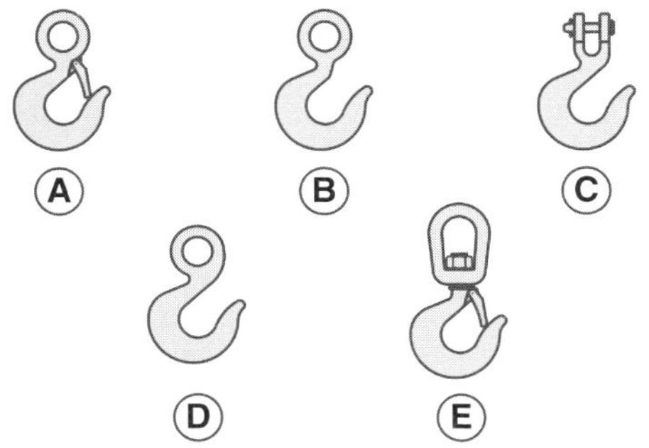

Rigging Hardware Attachments

_____ 1. Shackle

_____ 2. Eyebolt

_____ 3. Chain

_____ 4. Hook

_____ 5. Thimble

_____ 6. Rope

_____ 7. Clip

_____ 8. Choker fitting

_____ 9. Webbing

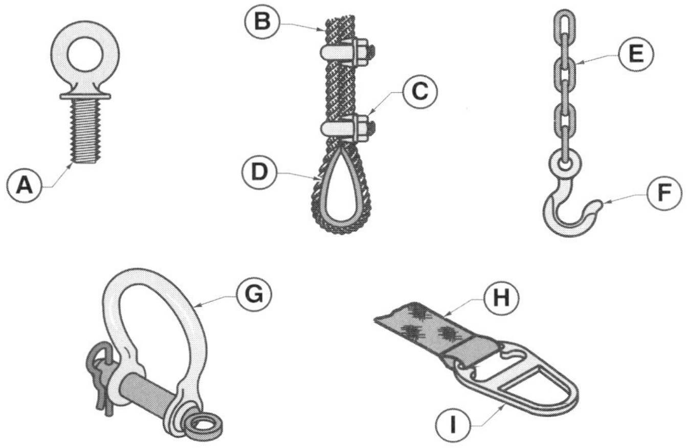

Wire Rope Terminations

_____ 1. Wedge socket

_____ 2. Open speltered socket

_____ 3. Closed speltered socket

_____ 4. Closed swaged socket

_____ 5. Thimble

_____ 6. Thimble and link

_____ 7. Thimble and shackle

_____ 8. Thimble and hook

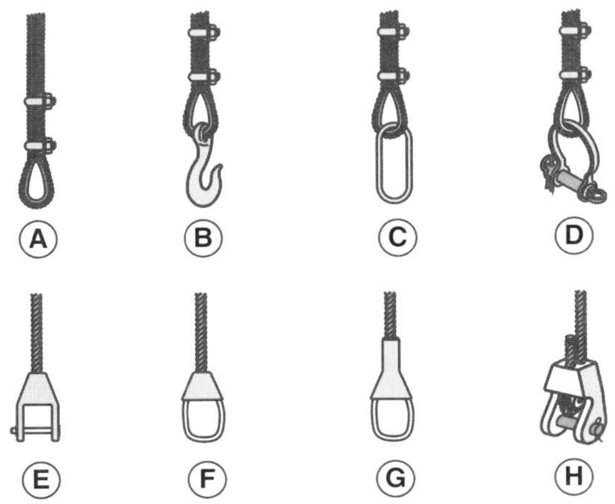

36 INDUSTRIAL MECHANICS WORKBOOK

Chain Inspection

_____ 1. Bent links

_____ 2. Cracks

_____ 3. Stretching

_____ 4. Excessive wear

_____ 5. Gouges

Problems

Refer to appropriate tables in Appendix.

_____ 1. The pieces of 1¾" D round steel weigh ___ lb.

_____ 2. Three 10' pieces of ¼" D round steel and one 10' piece of 1" square steel weighs ___ lb.

_____ 3. An order for twenty-four 36" × 96" sheets of ¼" steel plate weighs ___ lb.

_____ 4. The loss factor is ___ if the sling angle is 50° from the horizon.

_____ 5. The total lifting capacity of a two-leg sling made of ¼", 6 × 19, IPS-FC wire rope with the sling loops constructed of swaged sockets and sling angles of 60° is ___ t.

_____ 6. The rope bending load rating of a ½" rope traveling over an 8" pulley with a load rating of 1800 lb is ___ lb.

_____ 7. The generally accepted safe wire rope strength to lift 5000 lb with a steady lift is ___.

_____ 8. The lifting capacity of a basket hitch using a 1½" wide Class 5, Type V endless sling without fittings and having a 40° sling angle is ___ lb.

_____ 9. The lifting capacity of a round sling basket hitch with a yellow cover and 50° sling angle is ___ lb.

_____ 10. Should the used chain be removed from service?

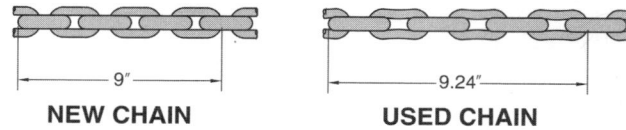

Rigging 5

TEST 2

Industrial Mechanics

_____ 1. ___ is hoisting equipment or machinery by mechanical means.

_____ 2. A(n) ___ load is a load in which one-half of the load is not a mirror image of the other half.

_____ 3. The ___ is the balancing point of a load.

_____ 4. The ___ weight center is a weight mass above a pivot point that causes a load to topple because it is too heavy.

_____ 5. A(n) ___ is a line consisting of a strap, chain, or rope used to lift, lower, or carry a load.

_____ 6. Fiber rope is constructed by twisting ___.
   A. fibers into yarn
   B. yarn into strands
   C. strands into rope
   D. all of the above

_____ 7. A(n) ___ is a bitter substance with a pH value greater than 7.

_____ 8. ___ is the wrapping placed around all strands of a rope near the area where the rope is cut.

T   F   9. A hitch is the interlacing of rope to temporarily secure it without knotting the rope.

T   F   10. A bowline knot is a knot that forms a loop which slips along the rope from which it is made.

T   F   11. A wagoneer's hitch knot is a knot that creates a load-securing loop from the standing part of the rope.

_____ 12. A timber hitch is ___.
   A. a binding knot and hitch combination
   B. used to wrap and drag lengthy material
   C. either A or B
   D. none of the above

38 INDUSTRIAL MECHANICS WORKBOOK

_____ 13. ___ strength is a measure of the greatest amount of straight-pull stress metal can bear without tearing apart.

_____ 14. ___ strength is a metal's resistance to deflection in the direction in which the load is applied.

_____ 15. The diameter of wire rope is determined by the largest possible ___ dimension.

_____ 16. ___ wire is wire rope that uses different size wire in different layers.

_____ 17. ___ is a rope's attempt to rotate and untwist its strand lays while under stress.

_____ 18. A(n) ___ is a complete helical wrap of the strands of a rope.

_____ 19. A(n) ___ is a rope splice containing a thimble.

_____ 20. The ___ of a web sling is a length of webbing folded back and spliced to the sling body, forming an opening.

_____ 21. A(n) ___ pad is a leather or webbed pad used to protect the web sling from damage.

_____ 22. A(n) ___ sling is a sling consisting of one or more continuous polyester fiber yarns wound together to make a core.

_____ 23. A(n) ___ is a series of metal rings connected to one another and used for support, restraint, or transmission of mechanical power.

_____ 24. Regarding wire ropes, a strength safety factor of ___ is used for shock or uneven loads.

_____ 25. Basic web slings are fabricated in six configurations, which are ___ I through VI.

Rope Lay

_____ 1. A(n) ___-lay is shown at A.

_____ 2. A(n) ___-lay is shown at B.

_____ 3. A(n) ___-lay is shown at C.

_____ 4. A(n) ___-lay is shown at D.

① STRANDS TWISTED IN CLOCKWISE ROTATION
② STRANDS TWISTED IN COUNTERCLOCKWISE ROTATION
③ YARN OR WIRES TWISTED IN CLOCKWISE ROTATION
④ YARN OR WIRES TWISTED IN COUNTERCLOCKWISE ROTATION

Rope Terminology

_____ 1. Loop
_____ 2. Kink
_____ 3. Standing part
_____ 4. Standing end
_____ 5. Working part
_____ 6. Whipping
_____ 7. Bight
_____ 8. Nip
_____ 9. Eye loop
_____ 10. Working end

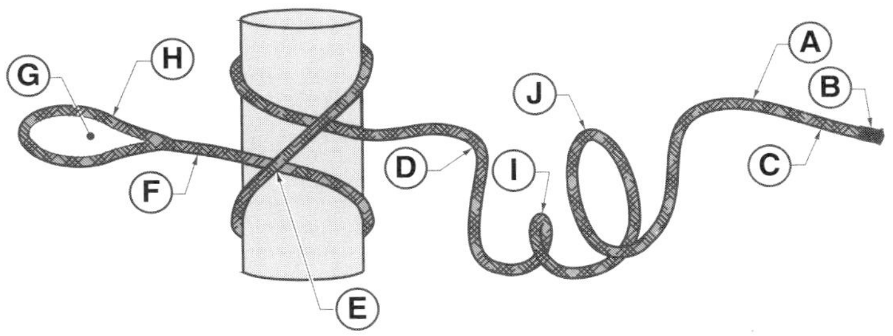

Hoisting Hooks

_____ 1. Foundry
_____ 2. Choker
_____ 3. Swivel
_____ 4. Grab
_____ 5. Sorting

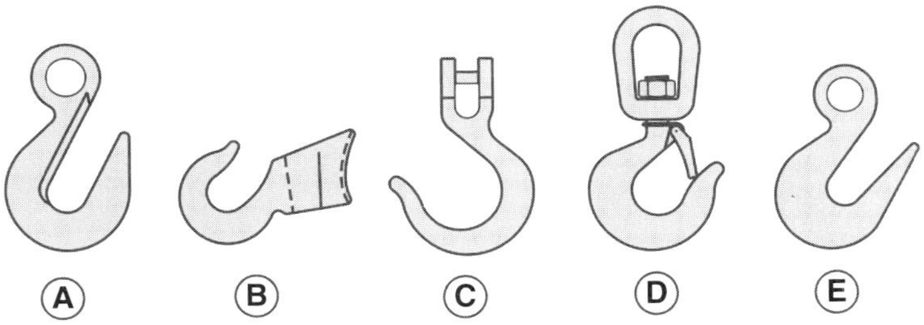

Basic Sling Combinations

_____ 1. Basket
_____ 2. Bridle
_____ 3. Choker
_____ 4. U
_____ 5. Vertical (single-leg)

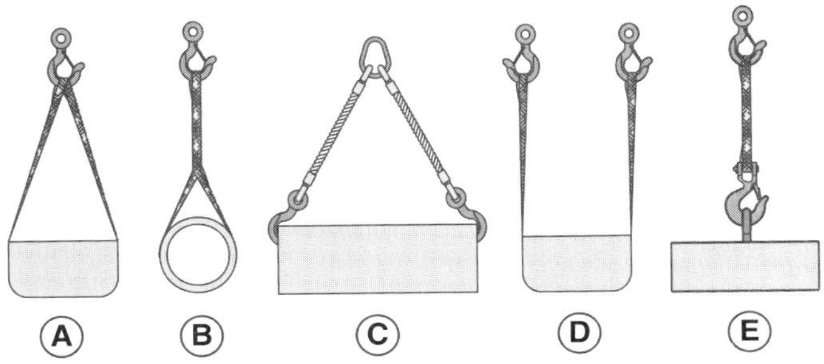

Slings

_____ 1. Type I

_____ 2. Type II

_____ 3. Type III

_____ 4. Type IV

_____ 5. Type V

_____ 6. Type VI

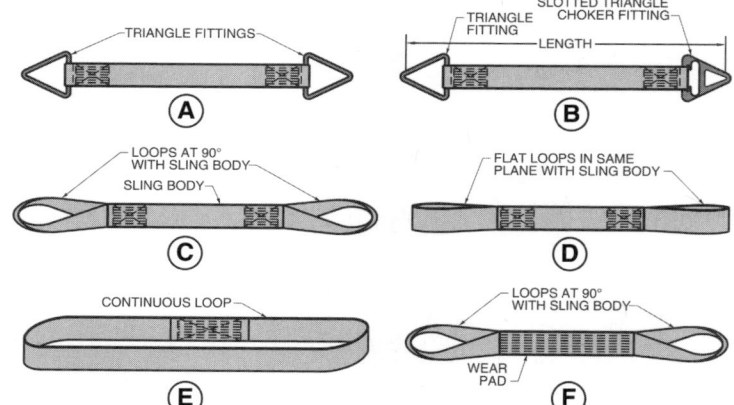

Problems

Refer to appropriate tables in Appendix.

_____ 1. The 10′ pieces of 1¼″ D brass weigh ___ lb.

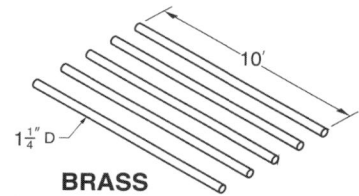

BRASS

_____ 2. Fifteen 10′ pieces of ¾″ D round steel and four 10′ pieces of 1″ square steel weigh ___ lb.

_____ 3. Twenty-eight 48″ × 96″ sheets of ¼″ steel plate weighs ___ lb.

_____ 4. The loss factor is ___ if the sling angle is 65° from the horizon.

_____ 5. The total lifting capacity of a two-leg sling made of ⅜″, 6 × 19, IPS-FC wire rope with the sling loops constructed of wedged sockets and sling angles of 70° is ___ t.

_____ 6. The rope bending load rating of a ½″ rope traveling over a 6″ pulley with a load rating of 1500 lb is ___ lb.

_____ 7. The generally accepted safe wire rope strength to lift 3000 lb with a steady lift is ___.

_____ 8. The lifting capacity of a basket hitch using a 1½″ wide Class 5, Type V endless sling without fittings and having a 50° sling angle is ___ lb.

_____ 9. The lifting capacity of a round sling basket hitch with a tan cover and 45° sling angle is ___ lb.

_____ 10. Should the used chain be removed from service?

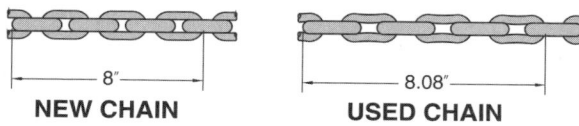

NEW CHAIN   **USED CHAIN**

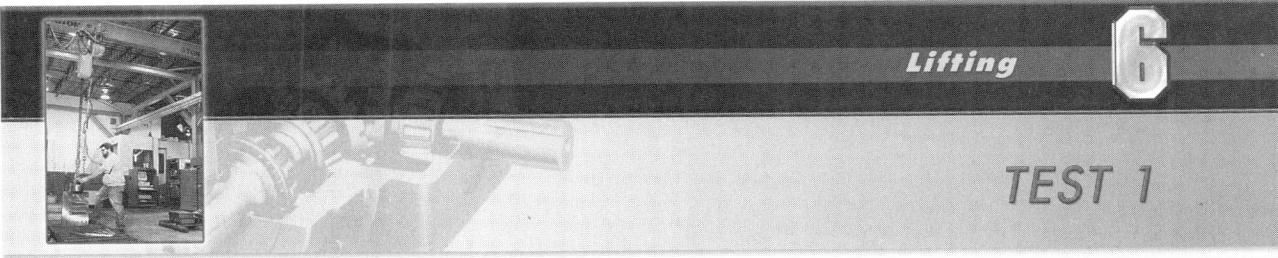

Lifting 6

TEST 1

Name _____ Date _____

Industrial Mechanics

_____ 1. ___ is the hoisting of equipment or machinery by mechanical means.

_____ 2. A ___ is a rope length between the lower block and the upper block of a block and tackle.
   A. piece
   B. part
   C. portion
   D. none of the above

_____ 3. A ___ line is the part of the rope to which force is applied to hold or move a load.
   A. load
   B. lead
   C. front
   D. back

_____ 4. Mechanical ___ is the ratio of the output force of a device to the input force.

T   F   5. The nominal bending strength of the most heavily loaded rope in a system shall be no less than 2½ times the load applied to that rope.

T   F   6. Torque is the twisting (rotational) force of a shaft.

T   F   7. In a bevel gear, the drive gear is the smaller gear.

T   F   8. Ambient temperature is the temperature of the air surrounding a piece of equipment.

_____ 9. A(n) ___ is a pushbutton or lever control suspended from a crane or hoisting apparatus.

_____ 10. The proper direction for winding the first layer of rope on a drum is determined by the ___ of the rope.
   A. length
   B. diameter
   C. lay
   D. none of the above

41

_____ **11.** The two basic types of eyebolts are formed steel and ___ steel.

T F **12.** As a sling moves from a vertical to an angular position, the capacity of the eyebolt is increased.

T F **13.** All crane pulls should be vertical.

_____ **14.** A(n) ___ is a bolt with a looped head.

_____ **15.** A(n) ___ is an assembly of hooks, pulleys, and frames suspended by hoisting ropes.

_____ **16.** ___ is passing a rope through a hole or opening or around a series of pulleys.

_____ **17.** A(n) ___ chain is the chain that raises the load.
   A. pull
   B. lift
   C. pickup
   D. hoist

_____ **18.** The ___ hook is the hook assembled to the top of a hoisting mechanism to allow for overhead suspension.
   A. main
   B. overhead
   C. top
   D. master

_____ **19.** Lever-operated hoists are generally used to lift loads that weigh from ___ lb to ___ lb.
   A. 100; 300
   B. 200; 500
   C. 300; 600
   D. none of the above

_____ **20.** A(n) ___ is a mechanism used to prevent the ratchet wheel of a lever-operated hoist from turning backwards.

_____ **21.** A(n) ___ hoist is a power-operated hoist operated by a geared reduction air motor.

_____ **22.** Drum ___ is the rope length required to make one complete turn around the drum of a hoist or crane.

_____ **23.** Hoist ___ is the slippage of a hook caused by insufficient braking.

_____ **24.** Typical eyebolt angular lift capacity is calculated using a constant of ___ for sling angles of less than 45°.

_____ **25.** A(n) ___ crane is a crane with bridge beams supported on legs.

Bevel Gear

_____ 1. Slip clutch

_____ 2. Hoist chain

_____ 3. Bearing

_____ 4. Drive gear

_____ 5. Endless hand chain

_____ 6. Beveled gears

_____ 7. Load

_____ 8. Driven gear

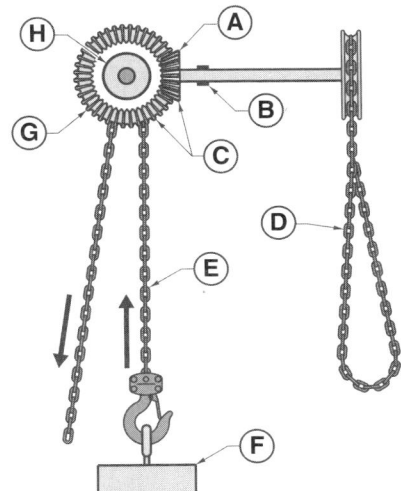

Drum Wrap

_____ 1. Underwind left to right

_____ 2. Overwind left to right

_____ 3. Underwind right to left

_____ 4. Overwind right to left

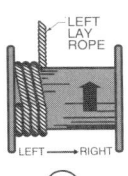 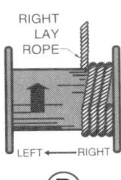 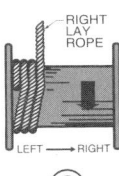 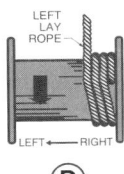

Safety

_____ 1. OSHA

_____ 2. ANSI

_____ 3. ISO

_____ 4. CMAA

_____ 5. ASME

_____ 6. NFPA

A. Publishes the National Electrical Code®, which contains standards for the practical safeguarding of persons and property from the hazards arising from the use of electricity.

B. Nongovernmental international organization comprised of national standards institutions of over 90 countries.

C. U.S. standards-developing organization that adopts and co-publishes standards that are written and approved by member organizations.

D. U.S. government organization concerned with the development and enforcement of safety standards for industrial workers.

E. Oranization of crane manufacturers that promotes standardiza-tion and establishes crane-operating practice standards.

F. Organization that helps establish safe structural design of hoists and cranes and sets safety standards.

Crane Hand Signals

_____ 1. Raise boom
_____ 2. Lower boom
_____ 3. Raise boom and lower load
_____ 4. Lower boom and raise load
_____ 5. Trolley travel
_____ 6. Multiple trolleys
_____ 7. Stop
_____ 8. Emergency stop
_____ 9. Move slowly
_____ 10. Bridge travel
_____ 11. Hoist
_____ 12. Lower
_____ 13. Use main hoist
_____ 14. Use auxiliary hoist

Hand-Chain Hoists

_____ 1. Hoist chain
_____ 2. Lower limit of hoist hook travel
_____ 3. Upper limit of hoist hook travel
_____ 4. Hoist hook
_____ 5. Reach
_____ 6. Pocket wheel
_____ 7. Top hook
_____ 8. Hand chain
_____ 9. Head room
_____ 10. Lift
_____ 11. Hand chain drop

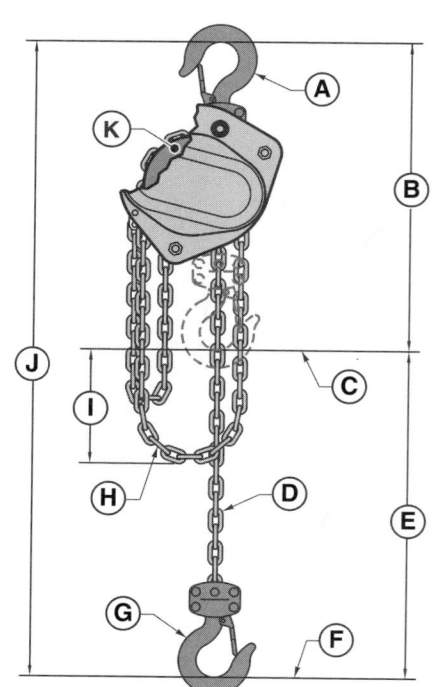

Jib Cranes

_____ 1. Wall-mounted, top-braced

_____ 2. Wall-mounted, cantilevered

_____ 3. Mast, cantilevered

_____ 4. Mast, underbraced

_____ 5. Mast, top-braced

_____ 6. Base-mounted, cantilevered

Problems

_____ 1. A force of ___ lb is required to hold a 600 lb load using a three-part reeving system.

_____ 2. Under ideal conditions, the lead line at A requires a pull of ___ ′ to lift the 100 lb load.

_____ 3. Under ideal conditions, the lead line at A requires a force of ___ lb to lift the 100 lb load.

_____ 4. A force of ___ lb is required to move an 8000 lb load using a 10-part reeving system equipped with rolling-contact bearing pulleys.

_____ 5. A(n) ___ lb force is required to move a 1000 lb load using one-part reeving and plain bearing pulleys.

_____ 6. The minimum compressor size required for a pneumatic hoist that requires 110 scfm is ___ HP.

_____ 7. The working load capacity of a 40° bridle sling using a ⅝″ shoulder nut eyebolt is ___ lb.

_____ 8. The working load of a 60° bridle sling using a ¼″ shoulder nut eyebolt is ___ lb.

_____ 9. The minimum output of a compressor rated at 11 HP is ___ scfm.

_____ 10. The minimum output of a compressor rated at 7.5 HP is ___ scfm.

_____ 11. The percent change in eyebolt capacity of a sling that has been relocated from 81° to 58° is ___%.

_____ 12. A force of ___ lb is required to move a 30,200 lb load using a four-part reeving system with rolling-contact bearing pulleys.

_____ **13.** A force of ___ lb is required to move a 21,925 lb load using a 15-part reeving system with plain bearing pulleys.

_____ **14.** A shim thickness of ___″ is required for a 90° rotation of a 1¼″ eyebolt.

_____ **15.** The working load capacity of a 40° bridle sling using a ½″ shoulder nut eyebolt is ___ lb.

Ladders and Scaffolds — TEST 1

Name _____ Date _____

Industrial Mechanics

_____ **1.** A heavy-duty, industrial, 250 lb capacity ladder has a Type ___ rating.
   A. IA
   B. I
   C. II
   D. III

_____ **2.** Fixed ladders are installed in a preferred pitch range between ___° and 90° from horizontal.
   A. 45
   B. 60
   C. 75
   D. none of the above

_____ **3.** Metal ladders should not be used within ___′ of electrical circuits or equipment.

_____ **4.** Most metal ladders are constructed of ___, which is a relatively light metal.

T   F   **5.** Fiberglass ladders conduct electricity when dry.

T   F   **6.** A fixed ladder is permanently attached to a structure.

_____ **7.** A(n) ___ ladder is an adjustable-height ladder with a fixed bed section and sliding, lockable fly section(s).

_____ **8.** All scaffolds ___′ or more above ground must have guardrails, midrails, and toeboards.

_____ **9.** Guardrails on scaffolds must be installed no less than ___″ or more than ___″ high, with a midrail.
   A. 24; 30
   B. 30; 36
   C. 38; 45
   D. 42; 48

T   F   **10.** The minimum netting mesh size for bodily fall protection is normally 6″ × 6″.

T   F   **11.** Border rope for safety nets shall have a 2500 lb breaking strength when new.

48 INDUSTRIAL MECHANICS WORKBOOK

_____ **12.** Nails smaller than ___d common must not be used to construct scaffolds.
        A. 8
        B. 10
        C. 12
        D. 16

_____ **13.** A safety net must be used anywhere a person is working ___′ or more above ground, water, machinery, etc. when the worker is not otherwise protected by a safety belt, lifeline, or scaffolding.
        A. 10
        B. 25
        C. 40
        D. 60

T   F   **14.** A person should always face the ladder when ascending or descending.

T   F   **15.** Ladders are designed for use by only one person unless specifically designated otherwise.

_____ **16.** A(n) ___ scaffold is a scaffold supported by overhead wire ropes.

_____ **17.** The maximum working height of a hydraulic scissor lift scaffold is ___′.

_____ **18.** Scaffold platform planks consist of ___″ nominal structural planks.

_____ **19.** A(n) ___ scaffold is a wood scaffold with one or two sides firmly resting on the floor or ground.

_____ **20.** Stepladders are commonly ___ in length.
        A. 2′-0″ to 6′-0″
        B. 2′-0″ to 8′-0″
        C. 4′-0″ to 8′-0″
        D. 4′-0″ to 10′-0″

_____ **21.** The spacing between rungs of ladders, except for stepstools, shall be on ___″ centers ± ___″.
        A. 8; 1/8
        B. 8; 1/4
        C. 12; 1/8
        D. 12; 1/4

_____ **22.** The overlap of the fly section of a 42′ extension ladder shall be at least ___′.

_____ **23.** The ___ is the rope used for raising and lowering the fly sections of extension ladders.

_____ **24.** The minimum distance between the center of the rung of a fixed ladder to the building wall is ___″.

_____ **25.** A cage, well, or ladder safety system must be provided where a single length of climb on a fixed ladder is greater than 24′ but less than ___′.

_____ 26. Ladder ___ rating is the weight (in lb) a ladder is designed to support under normal use.

_____ 27. A Type ___ stepladder is designed for light-duty, household use.

T   F   28. Fixed ladders are commonly constructed of steel or aluminum.

T   F   29. Single ladders are of fixed length having only one section.

T   F   30. A mobile scaffold may be moved with a worker on the platform.

_____ 31. A scaffold over ___′ in height must be secured with a guyline.
    A. 10
    B. 15
    C. 20
    D. 25

_____ 32. Fixed ladders over ___′ in length must have a cage, well, or ladder safety system.
    A. 10
    B. 12
    C. 24
    D. 30

_____ 33. The tip of a single or extension ladder must be at least ___′ above the roof line or top support.
    A. 1
    B. 2
    C. 3
    D. 4

_____ 34. Ladders over ___′ in height must be secured at the bottom.
    A. 8
    B. 10
    C. 12
    D. 15

_____ 35. For fixed ladders using a ladder safety system, rest platforms must be provided at maximum intervals of ___′.
    A. 100
    B. 150
    C. 175
    D. 200

_____ 36. The surface of a mobile scaffold must be within ___° of being level.
    A. 1½
    B. 2
    C. 2½
    D. 3

Sectional Metal-Framed Scaffolds

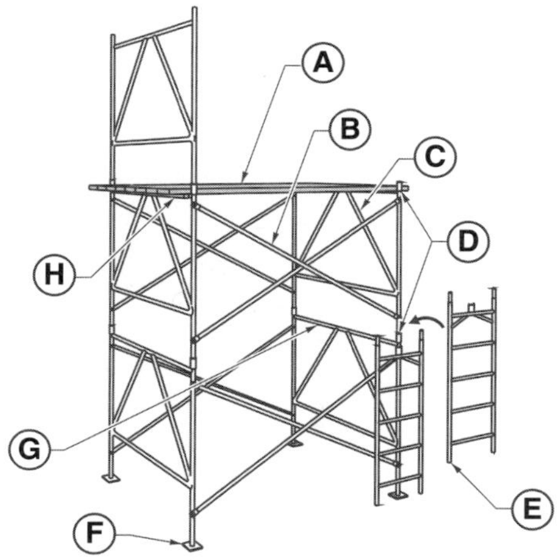

_____ 1. Bearer

_____ 2. Hook-on ladder

_____ 3. Diagonal brace

_____ 4. Cross brace

_____ 5. Coupling tube

_____ 6. Footing base plate

_____ 7. Cleat

_____ 8. Planking

Ladder Jacks

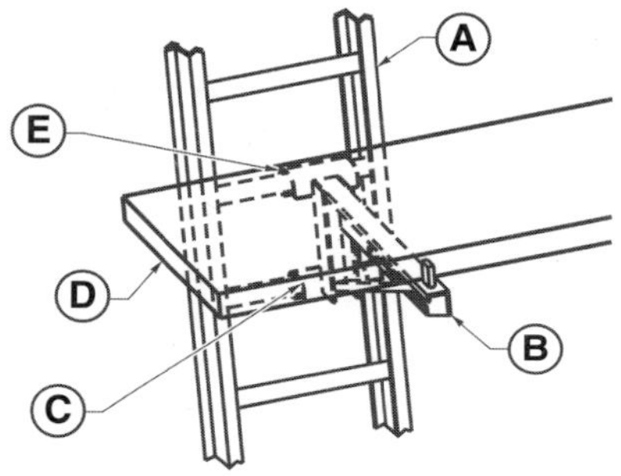

_____ 1. Hook

_____ 2. Ladder jack

_____ 3. Ladder

_____ 4. Plank

_____ 5. Hook

Pole Scaffolds

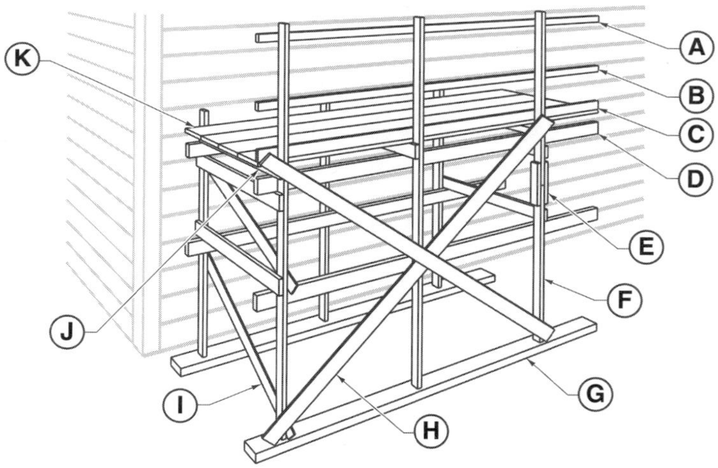

_____ 1. Ledger

_____ 2. Cross brace

_____ 3. Diagonal brace

_____ 4. Planking

_____ 5. Footing

_____ 6. Guardrail

_____ 7. Upright

_____ 8. Midrail

_____ 9. Toeboard

_____ 10. Bearer

_____ 11. Splice

Extension Ladders

_____ 1. Halyard

_____ 2. Rungs

_____ 3. Tip

_____ 4. Butt end

_____ 5. Plastic rail closures

_____ 6. Center swivel pulley

_____ 7. Foot assembly

_____ 8. Flange

_____ 9. Web

_____ 10. Bed section

_____ 11. Pawl lock

_____ 12. Fly section

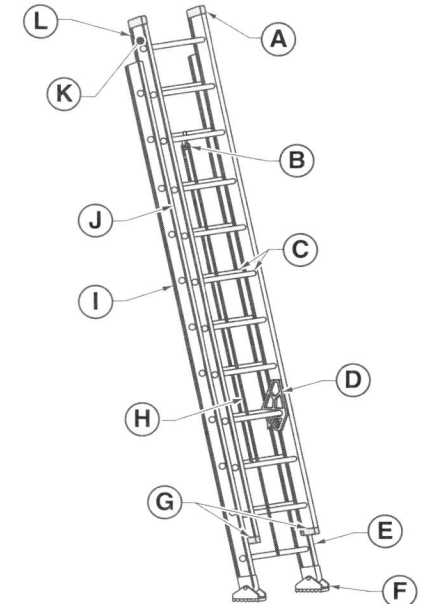

Pawl Locks

_____ 1. The fly section of the ladder at ___ is held in place.

_____ 2. The fly section of the ladder at ___ is being lowered.

_____ 3. The fly section of the ladder at ___ is being raised.

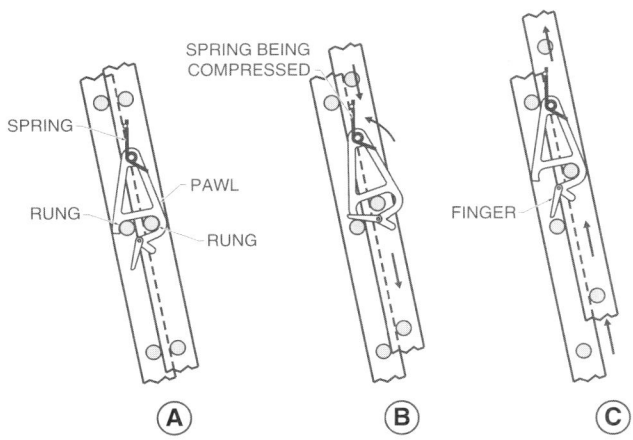

Problems

_____ 1. An extension ladder has a working height of 16′. The butt end of the ladder is placed ___′ from the wall.

_____ 2. A 54′ extension ladder shall have an overlap of at least ___′.

_____ 3. The braces of a medium-duty single-pole scaffold shall be constructed of ___ material.

_____ 4. Rails for light-duty double-pole scaffolds shall be constructed of ___ material.

_____ 5. The base of a metal-framed scaffold measures 6′ × 12′. The maximum height of the scaffold is ___′.

_____ 6. The minimum base dimension of the Mobile Scaffold is ___′.

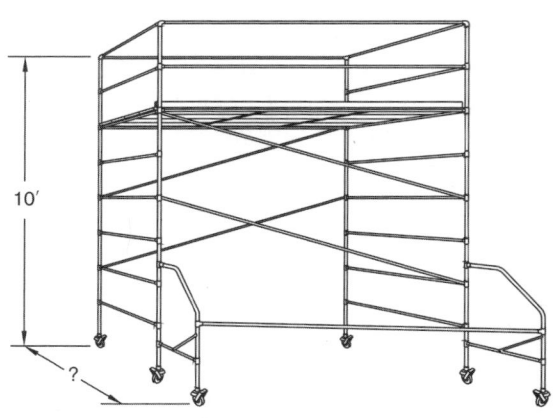

MOBILE SCAFFOLD

Hydraulic Principles 8

TEST 1

Name _____ Date _____

Industrial Mechanics

_____ 1. ___ is the branch of science that deals with the practical application of water or other liquids at rest or in motion.

_____ 2. ___ force is the outward force produced by a rotating object.

_____ 3. A(n) ___ is a fluid that has neither independent shape nor volume and tends to expand indefinitely.

_____ 4. ___ is the force per unit area.

_____ 5. ___ is the energy that produces movement.

_____ 6. The weight of the atmosphere at sea level is ___ psia.

_____ 7. A mercury barometer is commonly calibrated in ___.

_____ 8. The area of a circle is ___% of the area of a square with the same measurement.

T   F   9. Area, force, and pressure are the basis of all hydraulic systems.

T   F   10. The pressure of the fluid in a vessel is the same at that level regardless of the shape of the vessel.

T   F   11. Fluids that are thin and flow easily have a high viscosity.

T   F   12. One gallon of fluid equals 321 cu in.

T   F   13. The velocity of a fluid decreases as the cross-sectional area of a pipe increases.

_____ 14. A(n) ___ is a quantity that has a magnitude and direction.

_____ 15. Mechanical ___ is the ratio of the output force of a device to the input force.

_____ 16. ___ is a measure of the ability to do work.

_____ 17. ___ is the energy used when a force is exerted over a distance.

T   F   18. Mineral-based oil is the most widely-used hydraulic fluid.

T   F   19. Static head pressure is potential energy.

54 INDUSTRIAL MECHANICS WORKBOOK

_____ **20.** One horsepower is the amount of energy required to lift ___ lb 1′ in 1 min.
　　A. 330
　　B. 550
　　C. 33,000
　　D. 55,000

_____ **21.** ___ is the height at which atmospheric pressure forces a fluid above the elevation of its supply source.
　　A. Increase
　　B. Elevation
　　C. Lift
　　D. none of the above

_____ **22.** In a hydraulic system, ___.
　　A. pressure provides force
　　B. flow rate provides speed
　　C. both A and B
　　D. neither A nor B

T　F　**23.** Volume is the two-dimensional size of an object measured in cubic units.

T　F　**24.** Flow rate is the volume of fluid flow.

T　F　**25.** Static energy is the energy of motion.

Fluid Pressure

T　F　**1.** The pressure at A is twice the pressure at D.

T　F　**2.** The pressure at B is the same as the pressure at C.

T　F　**3.** The pressure at E is greater than the pressure at D.

T　F　**4.** The pressure at F is the same as the pressure at D.

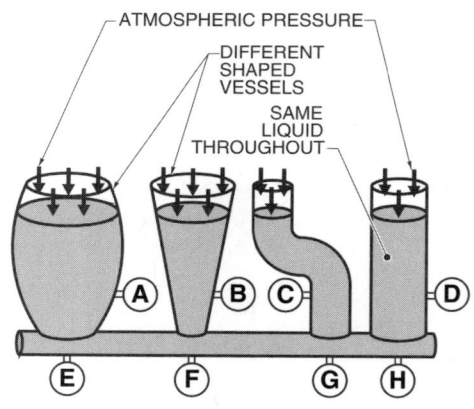

Lift

_____ 1. ___ lift is shown at Tank A.

_____ 2. ___ lift is shown at Tank B.

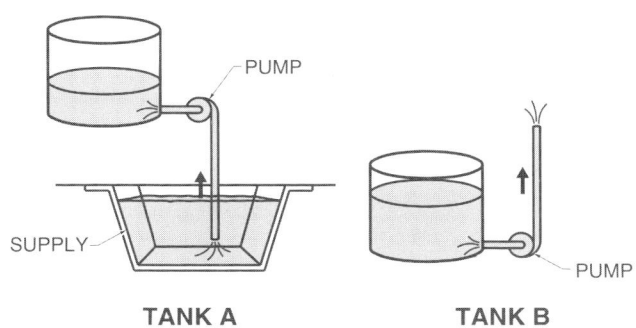

Fluid Flow

_____ 1. The fluid velocity at A is ___ ft/sec. (Velocity is 4x greater in a pipe of ½ dia.)

_____ 2. The fluid velocity at B is ___ ft/sec. (Velocity is 4x greater in a pipe of ½ dia.)

    T   F 3. The fluid velocity at C is ¼ the fluid velocity at A. (Velocity is 4x greater in a pipe of ½ dia.)

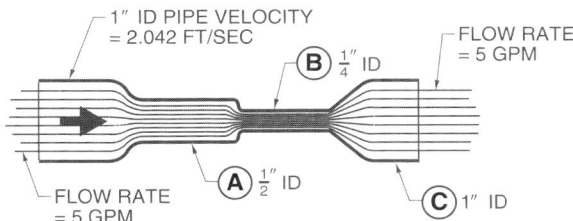

Horsepower

_____ 1. The horsepower required to lift Load A is ___ HP.

_____ 2. The horsepower required to lift Load B is ___ HP.

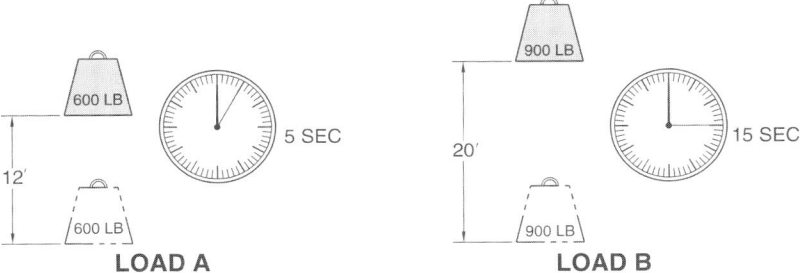

Problems

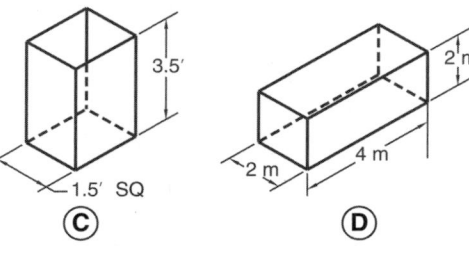

_____ 1. Tank C has a capacity of ___ cu ft.

_____ 2. Tank D has a capacity of ___ m³.

_____ 3. The absolute pressure in a system with a gauge pressure of 94.5 psig is ___ psia.

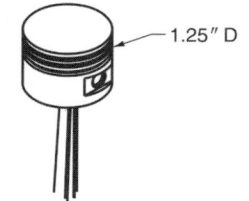

_____ 4. The area of Piston A is ___ sq in.

_____ 5. A pressure of ___ psi is required to move a 1000 lb force with Piston A.

_____ 6. The amount of fluid required to fully extend a 3.5″ D cylinder with an 18″ stroke is ___ gal.

_____ 7. A force of ___ lb is produced by a 4 sq in. piston operating at 125 psi.

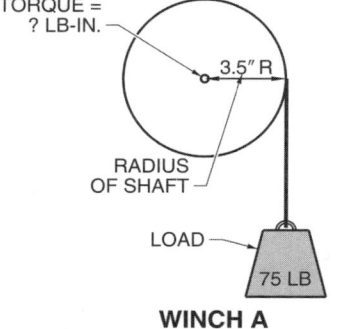

PISTON A

_____ 8. The velocity of a fluid having a flow rate of 4.5 gpm through a 1′ section of ¾″ D pipe is ___ ft/sec.

_____ 9. The torque required to overcome the force at Winch A is ___ lb-in.

_____ 10. If the distance in Problem 9 was increased to 5.25″, ___ lb-in would be required to overcome the force.

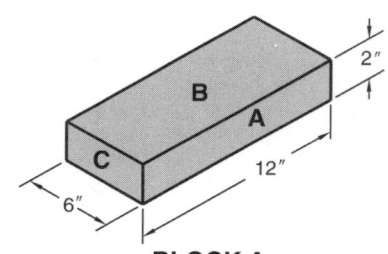

WINCH A

_____ 11. The area of Surface A on Block A is ___ sq in.

_____ 12. The area of Surface B on Block A is ___ sq ft.

_____ 13. The volume of Block A is ___ cu in.

_____ 14. The velocity in Pipe A is ___ times greater than the velocity in Pipe B.

_____ 15. The area of Pipe A is ___ sq in.

BLOCK A

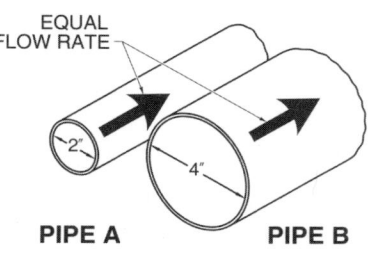

PIPE A   **PIPE B**

Hydraulic Principles

TEST 2

Industrial Mechanics

_____ 1. ___ is the study of liquids at rest and the forces exerted on them or by them.

_____ 2. ___ is the condition when all forces and torques are balanced by equal and opposite forces and torques.

_____ 3. A mercury ___ is an instrument that measures atmospheric pressure using a column of mercury.

_____ 4. ___ is a pressure lower than atmospheric pressure.

_____ 5. Area is always expressed in ___ units.
   A. square
   B. cubic
   C. either A or B
   D. neither A nor B

_____ 6. ___ lift is the lift of fluid in motion.
   A. Static
   B. Dynamic
   C. Head
   D. none of the above

T   F   7. Any friction generated in a hydraulic system becomes a resistance to fluid flow.

T   F   8. The velocity of a fluid is constant as its speed or direction of flow changes from one moment to another.

T   F   9. Total energy is a measure of a fluid's ability to do work.

_____ 10. ___ is the distance a fluid travels in a specified time.

_____ 11. A(n) ___ is a support on which a lever turns or pivots and is located somewhere between the effort force and the resistance force.

_____ 12. ___ energy is the energy of motion.

_____ 13. ___ is the twisting (rotational) force of a shaft.

_____ 14. ___ is the study of the forces exerted on a solid body by the motion or pressure of a fluid.

_____ 15. The ___ pressure is pressure above a perfect vacuum.

T   F   16. Head is the difference in the level of a liquid between two points.

T   F   17. Fluids that flow with difficulty have a low viscosity.

T   F   18. Capacity is expressed in square units.

_____ 19. ___ torque is the energy that a motor develops to keep a load turning.

_____ 20. One horsepower equals ___ lb ft/sec.

_____ 21. ___ is the rate or speed of doing work.
   A. Energy
   B. Power
   C. Capacity
   D. Efficiency

_____ 22. The viscosity ___ is a measure of the degree to which viscosity changes when a fluid is heated.
   A. rate
   B. index
   C. time
   D. temperature

_____ 23. Fluid ___ is the movement of fluid caused by a difference in pressure between two points.

_____ 24. ___ lift is the height to which atmospheric pressure causes a column of fluid to rise above the supply to restore equilibrium.

_____ 25. ___ pressure is pressure above atmospheric pressure that is used to express pressures inside a closed system.

_____ 26. A pressure gauge reads ___ psig at normal atmospheric pressure.

_____ 27. ___ is an increase in speed.

_____ 28. ___ head is the head of fluid in motion.
   A. Static
   B. Still
   C. Dynamic
   D. Divergent

_____ 29. ___ is the volume of oil moved during each cycle of a pump.
   A. Residue
   B. Displacement
   C. Load
   D. none of the above

_____ **30.** ___ is a measure of a component's or system's useful output energy.
   A. Rate
   B. Percentage
   C. Efficiency
   D. Value

_____ **31.** The amount of pressure required to move a 6800 lb force with a 6″ D piston is ___ psi.
   A. 240
   B. 240.5
   C. 241
   D. 241.5

_____ **32.** The area of a circle with a diameter of 4.275″ is ___ sq in.
   A. 4.57
   B. 14.35
   C. 16.08
   D. 18.28

_____ **33.** The absolute pressure in a system under standard conditions with a gauge reading of 212 psig is ___ psia.
   A. 212.49
   B. 213.47
   C. 226.70
   D. 241.92

_____ **34.** The pressure at the base of a 55 gal. polyethylene drum with an inside diameter of 23½″ and a height of 36¼″ filled with kerosene is ___ lb/sq in.
   A. 1.07
   B. 1.10
   C. 1.17
   D. 1.73

_____ **35.** The velocity of cold water flowing at 100 gpm through a 60″ section of 3″ D galvanized pipe is ___ ft/sec.
   A. 1.63
   B. 4.54
   C. 19.62
   D. 40.62

_____ **36.** The rod speed of a 9.1 gpm multistage booster pump with a 4″ D cylinder is ___ in./min.
   A. 18.38
   B. 94.31
   C. 167.23
   D. 320.83

60 INDUSTRIAL MECHANICS WORKBOOK

_____ **37.** The amount of horsepower required to move a 20,380 lb steel coil 100′ in 45 sec is ___ HP.
   A. 8.24
   B. 22.04
   C. 82.34
   D. 102.74

Hydrostatics

_____ **1.** Heat energy

_____ **2.** Static energy

_____ **3.** Kinetic energy

_____ **4.** External force

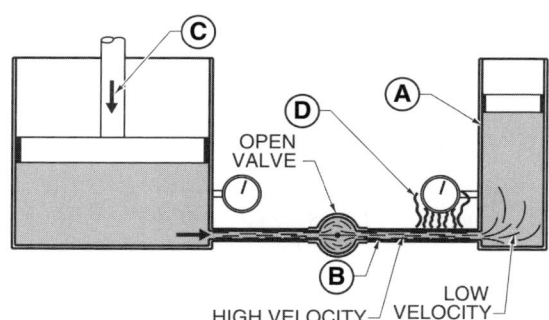

Mechanical Advantage

T   F   **1.** Gear B will turn in a clockwise direction.

T   F   **2.** Gear B will turn twice as fast as Gear A.

T   F   **3.** Gear B will turn with twice as much force as Gear A.

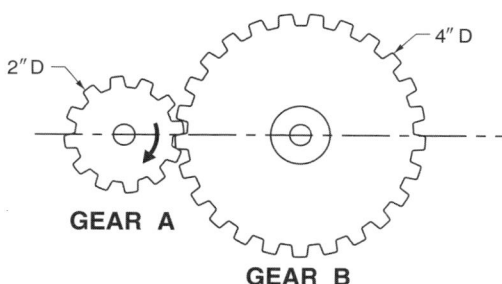

Problems

_____ 1. The volume of Cylinder A is ___ cu in.

_____ 2. The volume of Cylinder B is ___ mm³.

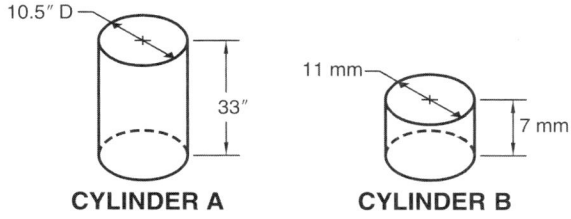

CYLINDER A   **CYLINDER B**

_____ 3. The area of Piston A is ___ mm².

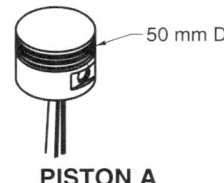

PISTON A

_____ 4. ___ gal. of fluid is required to fully retract the piston in Cylinder A.

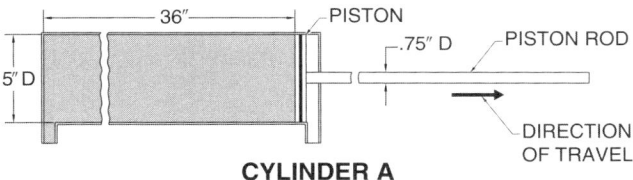

CYLINDER A

_____ 5. A fluid particle in a hydraulic system travels 78′ from 10:20:05 AM to 10:20:47 AM. The velocity of the fluid particle is ___ ft/sec.

_____ 6. ___ lb of effort force is required to lift the resistance force of Fulcrum A.

_____ 7. If the fulcrum in Fulcrum A were moved 2′ closer to F_2, and everything else remained the same, ___ lb of effort force would be required to lift F_2.

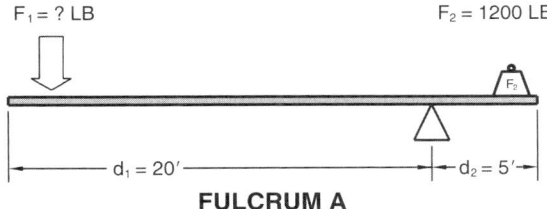

FULCRUM A

_____ 8. The torque required to overcome the force at Winch A is ___ lb-in.

_____ 9. If the load at Winch A was 120 lb, ___ lb-in would be required to overcome the force.

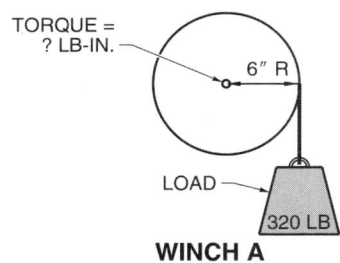

WINCH A

_____ 10. The horsepower required to lift the 3 t load is ___ HP.

_____ 11. If the 3 t load took 1 min to be lifted the 4′ distance, the horsepower required would be ___ HP.

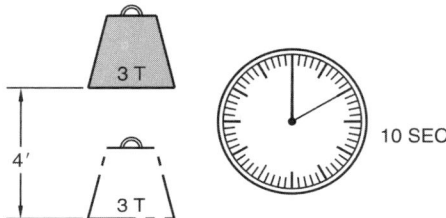

T   F   12. Load A requires more horsepower to be lifted than Load B.

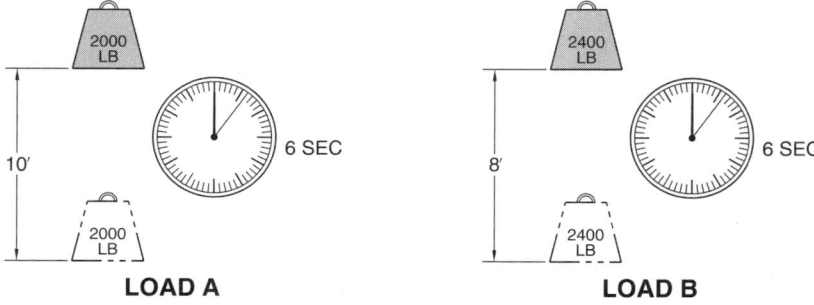

LOAD A   **LOAD B**

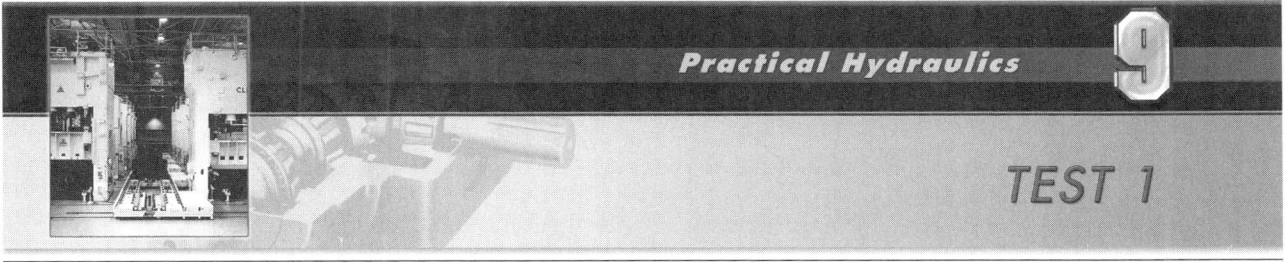

Practical Hydraulics — TEST 1

Name _____ Date _____

Industrial Mechanics

_____ 1. A(n) ___ is a fluid that can flow readily and assume the shape of its container.

_____ 2. A(n) ___ diagram shows the internal details of components and the path of fluid flow.

T F 3. Graphic symbols show flow paths, connections, and functions of components.

T F 4. Hydraulic fluids lubricate moving parts of a circuit.

_____ 5. ___ is the combining of oxygen with elements in oil which break down the basic oil composition.
  A. Foaming
  B. Pitting
  C. Oxidation
  D. Cavitation

_____ 6. ___ is localized corrosion that has the appearance of cavities.

T F 7. Strainer screens are rated in microns and filters are rated in mesh.

T F 8. A flared fitting is a fitting that is connected to a hose whose end is spread outward.

T F 9. The higher the mesh number of a strainer, the smaller the opening.

_____ 10. A hydraulic ___ is a device that converts hydraulic energy into straight-line (linear) energy.

_____ 11. For graphic symbols used in hydraulic circuits, ___ generally represent valves.

_____ 12. ___ is excessive air in hydraulic fluid.

_____ 13. The ___ is the temperature at which oil gives off enough gas vapor to ignite briefly when touched with a flame.

_____ 14. A(n) ___ valve is a valve that is activated or directly moved by a fluid pressure from the parallel port.

_____ 15. The ___ is the order in which a series of operations or movements are performed.

_____ 16. A ___ filter is positioned in a hydraulic circuit just before the reservoir.
        A. suction
        B. pressure
        C. return-line
        D. none of the above

_____ 17. A ___ is a device that transfers heat through a conducting wall from one fluid to another.
        A. fin cooler
        B. heat exchanger
        C. both A and B
        D. none of the above

_____ 18. A(n) ___ is a hollow cylinder of metal or other material of substantial wall thickness.

T   F   19. Tubing may be connected by welding or compression.

T   F   20. Positive displacement is the moving of a fixed amount of a substance with each cycle of a hydraulic pump.

T   F   21. A spur gear has straight teeth parallel to the shaft axes.

_____ 22. A(n) ___ valve is an infinite-position valve that has a disk that is raised or lowered over a port through which fluid flows.

_____ 23. A(n) ___ seal is used between moving parts to prevent leakage or contamination.

_____ 24. A(n) ___ motor is a device that converts hydraulic energy into mechanical energy.

_____ 25. ___ metals are metals containing iron.

_____ 26. A ___ diagram is the layout, plan, or sketch of a hydraulic circuit and is designed to explain, demonstrate, or clarify the relationship or functions between hydraulic components.
        A. pictorial
        B. cutaway
        C. hydraulic
        D. graphic

_____ 27. Without a ___, dismantling of equipment can be required to determine the function of components within a circuit.
        A. graphic diagram
        B. piston pump
        C. cutaway diagram
        D. cam ring

_____ 28. Hydraulic fluid is used as ___ because it can be applied instantly throughout a system and allows for control at different locations, increase or decrease of force, or a change of direction.
      A. resistance
      B. power
      C. force
      D. none of the above

_____ 29. A(n) ___ is a substance that causes harm or damage to that with which it comes in contact.
      A. additive
      B. contaminant
      C. lubricant
      D. lichen

_____ 30. ___ can be installed inside a reservoir at the fluid inlet of the system and placed in-line between the reservoir and the pump or installed on the exterior of a reservoir.
      A. Strainer screens
      B. Filters
      C. Suction strainers
      D. Pressure filters

_____ 31. Maximum drop value for filters is typically ___ psi.
      A. 30
      B. 45
      C. 60
      D. 75

_____ 32. Controlling pump ___ consumes the least horsepower while generating the least heat.
      A. input
      B. friction
      C. lubrication
      D. output

_____ 33. ___ is caused by low inlet pressure and is a change in size of the air molecules normally found in hydraulic fluids.
      A. Pseudocavitation
      B. Cavitation
      C. Pump discharge
      D. all of the above

_____ 34. Damage caused by ___ causes a seal to exhibit a hard and brittle material with cracks, and broken or chipped body or lip parts.
      A. heat
      B. chemicals
      C. contamination
      D. condensation

66 INDUSTRIAL MECHANICS WORKBOOK

_____ **35.** Other work-habit causes of contamination or damage to seals include ___.
　　A. improper handling of a seal before and during installation
　　B. installation of the wrong seal
　　C. installing a seal backwards
　　D. all of the above

Diagram Color Coding

_____ 1. Red

_____ 2. Yellow

_____ 3. Orange

_____ 4. Green

_____ 5. Blue

_____ 6. White

A. inactive fluid

B. intermediate presssure that is lower than system operating pressure

C. controlled flow by a metering device or lowest working pressure

D. exhaust or return flow to the reservoir

E. fluid flowing at system operating pressure or highest working pressure

F. intake flow to pump or drain line flow

Graphic Symbols — Lines

_____ 1. Main line

_____ 2. Pilot line

_____ 3. Drain line

_____ 4. Enclosure line

　　Ⓐ　　　　Ⓑ　　　　Ⓒ　　　　Ⓓ

Linear Equivalents

_____ 1. 1 in.

_____ 2. .001 mm

_____ 3. .0394 in.

_____ 4. 1μ

Ⓐ	25.4 mm	25,400μ
1 mm	Ⓑ	1000μ
1μ	25,400 of an in.	Ⓒ
Ⓓ	3.94 x 10⁻⁵ in.	.000039 in.

Graphic Symbols—Squares or Rectangles

_____ 1. Pressure switch

_____ 2. Single-acting cylinder

_____ 3. Double-acting cylinder

_____ 4. Directional valve

_____ 5. Pressure-relief valve

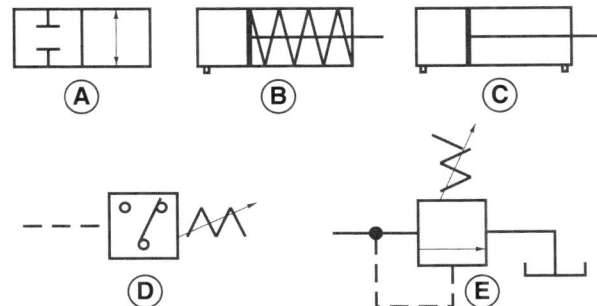

Gears

_____ 1. Herringbone

_____ 2. Helical

_____ 3. Spur

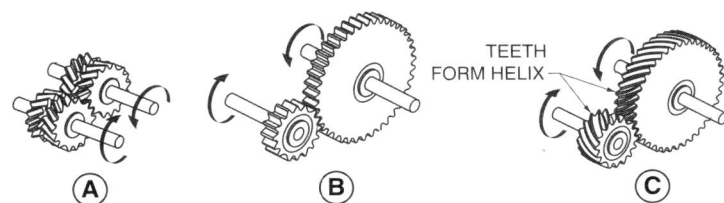

Pressure Gauge

_____ 1. Scale

_____ 2. Pivot

_____ 3. Spring

_____ 4. Pointer

_____ 5. Siphon connection

_____ 6. Pointer gear

_____ 7. Bourdon tube

_____ 8. Gear linkage

_____ 9. Linkage arm

_____ 10. Case

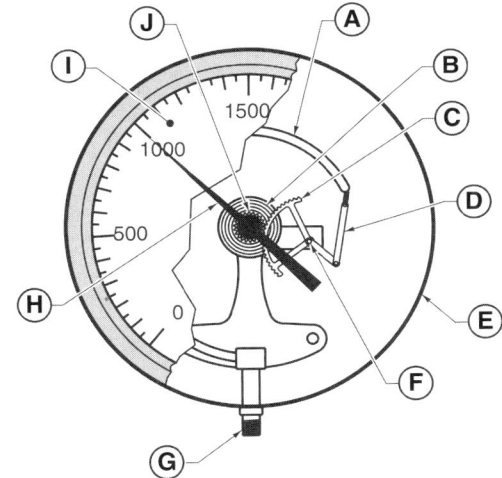

68 INDUSTRIAL MECHANICS WORKBOOK

Check Valves

_____ 1. Poppet

_____ 2. Ball

_____ 3. Pilot-operated

Problems

_____ 1. The minimum bending radius of Tubing A is ___″ R

TUBING A

2. Add a single-acting spring-return cylinder to Circuit A.

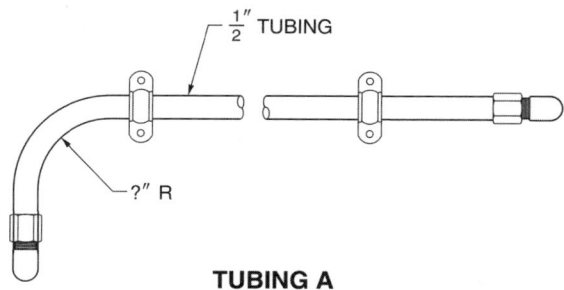

CIRCUIT A

3. Add a manually-actuated, spring-return, 3-way, two-position valve to control the cylinder in Circuit B.

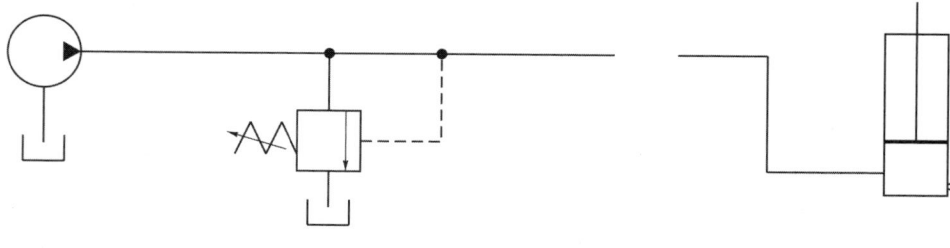

CIRCUIT B

Practical Hydraulics 9

TEST 2

Industrial Mechanics

_____ 1. ___ is the branch of science that deals with the practical application of water or other liquids at rest or in motion.

_____ 2. A(n) ___ is a closed path through which hydraulic fluid flows or may flow.

_____ 3. A(n) ___ diagram uses drawings or pictures to show the relationship of each component in a circuit.

_____ 4. A(n) ___ is a graphic element which indicates a particular device, etc.

_____ 5. ___ is the temperature at which oil ignites by itself.
   A. Flash point
   B. Fire point
   C. Auto-ignition
   D. none of the above

T   F   6. Strainer screens are rated in mesh and filters are rated in microns.

T   F   7. Pipe is designated according to its nominal size and wall thickness.

_____ 8. A(n) ___ is a container for storing fluid in a hydraulic system.

_____ 9. A(n) ___ is a mechanical device that causes fluid to flow.

_____ 10. ___ is the process in which microscopic gas bubbles expand in a vacuum and suddenly implode when entering a pressurized area.

_____ 11. A Bourdon tube is a hollow metal tube made of brass or similar material and is ___.
   A. elliptical in cross-sectional area
   B. bent in a C-shape
   C. both A and B
   D. none of the above

_____ 12. ___ is the capability of a material to regain its original shape after being bent, stretched, or compressed.
    A. Plasticity
    B. Revertance
    C. Resiliency
    D. none of the above

_____ 13. A(n) ___ is a seal used between machined parts or around pipe joints to prevent the escape of fluids.

_____ 14. ___ energy is the energy of motion.

_____ 15. For graphic symbols used in hydraulic circuits, ___ that are completely shaded generally represent liquid flow.

_____ 16. ___ are used in graphic diagrams to indicate an adjustable or variable component or to show shaft rotation on the near side of the shaft.
    A. Dashed lines
    B. Solid lines
    C. Dotted lines
    D. Arrows

_____ 17. A ___ is a device containing a porous substance through which a fluid can pass but particulate matter cannot.
    A. funnel
    B. strainer
    C. filter
    D. mask

_____ 18. A(n) ___ is a flexible tube for carrying fluids under pressure.

_____ 19. The standard flare angle for hydraulic tube fittings is ___° from the centerline.

T   F   20. Tubes should always be assembled in a straight line.

T   F   21. A ferrule is a metal sleeve used for joining one piece of tubing to another.

T   F   22. Vane pumps are the most widely used hydraulic pumps because of their simple design and ease of repair.

_____ 23. A(n) ___ is a device that controls the pressure, direction, or rate of fluid flow.

_____ 24. A hydraulic ___ is a device that converts hydraulic energy into mechanical energy.

_____ 25. The container in which fluid is stored under pressure in a hydraulic system is the ___.

Graphic Symbols — Circles

_____ 1. Pump
_____ 2. Motor
_____ 3. Pressure gauge
_____ 4. Flow meter
_____ 5. Check valve

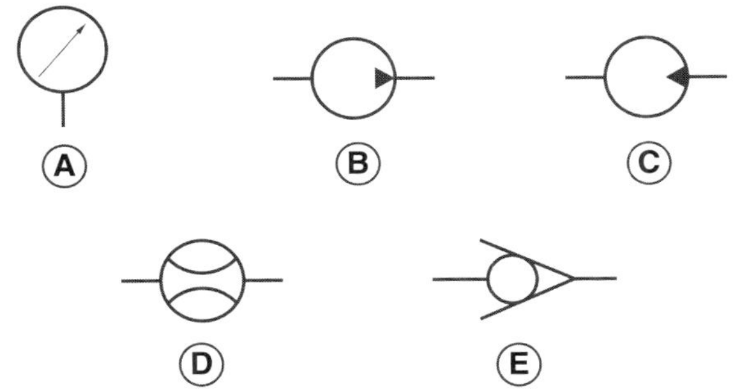

Graphic Symbols — Triangles

_____ 1. Motor
_____ 2. Air compressor
_____ 3. Bidirectional motor
_____ 4. Direction of flow

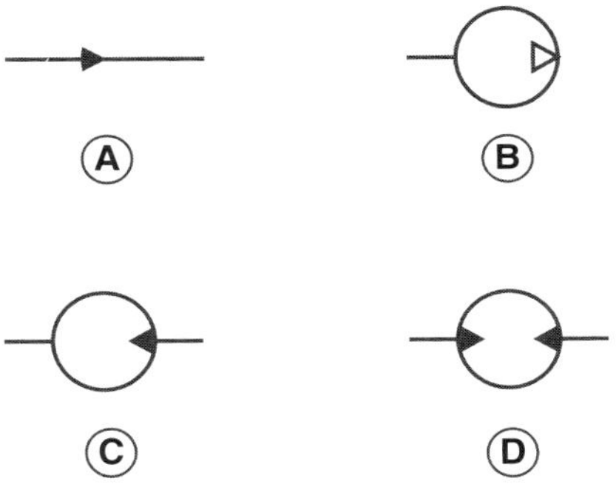

Flow Control Valves

_____ 1. Restrictive check valve
_____ 2. Globe
_____ 3. Gate
_____ 4. Needle

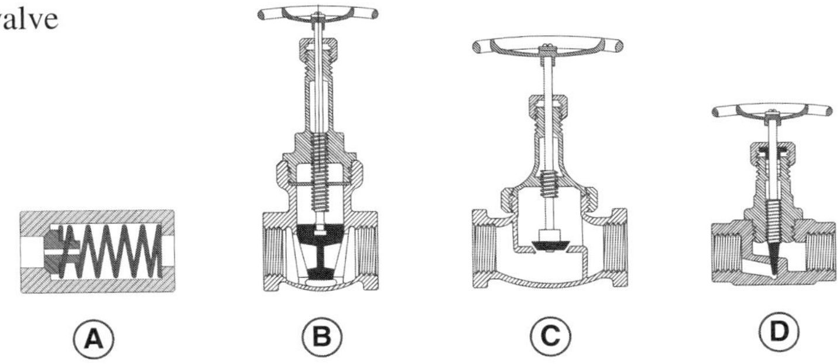

Filters

_____ 1. Pressure

_____ 2. Suction

_____ 3. Return-line

Ⓐ   Ⓑ   Ⓒ

Single Actuators

_____ 1. Manual

_____ 2. Pushbutton

_____ 3. Lever

_____ 4. Foot pedal

_____ 5. Solenoid

_____ 6. Mechanical

_____ 7. Detent

_____ 8. Air pilot

_____ 9. Spring

_____ 10. Oil pilot

Ⓐ Ⓑ Ⓒ Ⓓ

Ⓔ Ⓕ Ⓖ Ⓗ

Ⓘ Ⓙ

Problems

_____ 1. The minimum bending radius of Hose A is ___″ R.

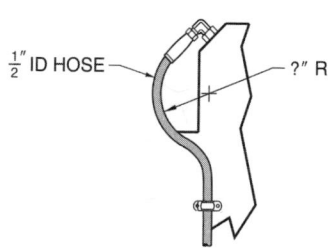

HOSE A

2. Add a pressure-relief valve to Circuit A.

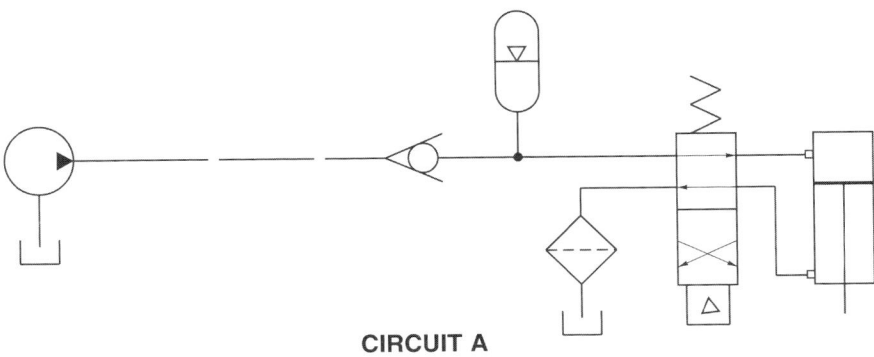

CIRCUIT A

3. Add a solenoid-actuated, spring-return, 4-way, two-position directional control valve to control the fluid flow in Circuit B.

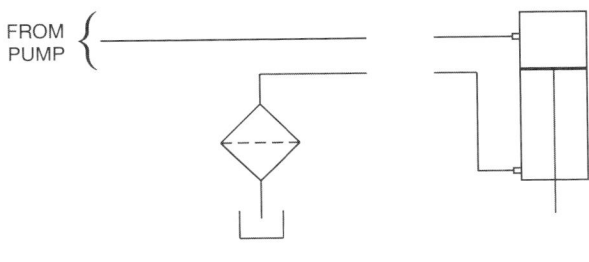

CIRCUIT B

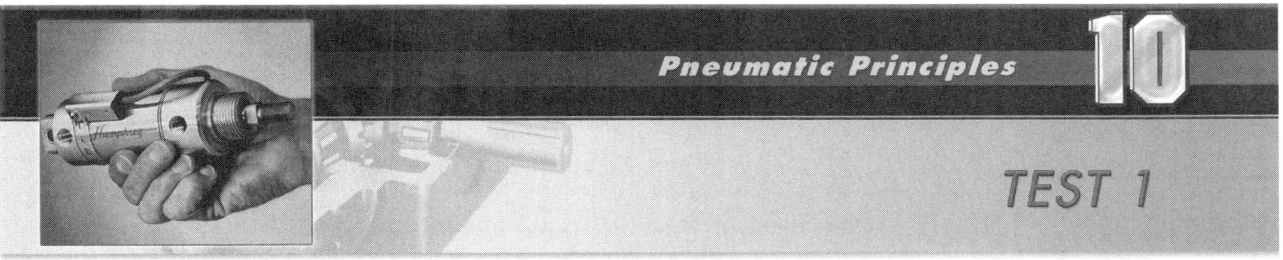

Pneumatic Principles 10

TEST 1

Name _____ Date _____

Industrial Mechanics

_____ 1. ___ is the branch of science that deals with the transmission of energy using a gas.

_____ 2. A(n) ___ is the smallest building block of matter than cannot be divided into smaller units without changing its basic character.

_____ 3. ___ is the force per unit area.

_____ 4. Atmospheric pressure at sea level is about ___ psia.

_____ 5. ___ is the three-dimensional size of an object measured in cubic units.

_____ 6. ___ pressure is pressure above a perfect vacuum.

_____ 7. A(n) ___ is pressure lower than atmospheric pressure.

_____ 8. Absolute ___ is the temperature at which substances possess no heat.

_____ 9. The temperature in °R is always ___° greater than the temperature in °F.
   A. 32
   B. 212
   C. 460
   D. 492

_____ 10. Free air is air at ___.
   A. atmospheric pressure
   B. ambient temperature
   C. both A and B
   D. neither A nor B

T F 11. Gas can expand to fill the volume and shape of its container.

T F 12. Gas molecules can be pushed closer together, allowing gas to be compressed.

T F 13. The pressure in a container varies as the size or shape of the container varies.

_____ 14. ___ pressure is the pressure above atmospheric pressure that is used to express pressures inside a closed system.

76 INDUSTRIAL MECHANICS WORKBOOK

_____ **15.** A(n) ___ compressor is a device that compresses gas by means of a piston(s) that moves back and forth in a cylinder.

_____ **16.** ___ is the amount of moisture in the air.

_____ **17.** A(n) ___ is a fine solid particle which remains individually dispersed in a gas.

_____ **18.** ___ is the change in state from a gas or vapor to a liquid.

_____ **19.** ___ is the area of industry that deals with the measurement, evaluation, and control of process variables.

_____ **20.** Atoms combine to form ___.
    A. protons
    B. particles
    C. molecules
    D. neither A, B, nor C

_____ **21.** The pressure exerted on Earth's surface varies with ___.
    A. altitude
    B. temperature
    C. humidity
    D. A, B, and C

T   F   **22.** In compression, air temperature decreases as a piston extends and the air molecules are forced closer together.

T   F   **23.** In an air compressor, multistage compression is required when the ratio of compression is greater than 6.

T   F   **24.** The total amount of moisture that air is capable of holding varies based on the temperature of the air.

_____ **25.** ___ air is air that holds as much moisture as it is capable of holding.

States of Matter

_____ **1.** Solid

_____ **2.** Liquid

_____ **3.** Gas

Gas Laws

_____ 1. Boyle's law

_____ 2. Charles' law

_____ 3. Gay-Lussac's law

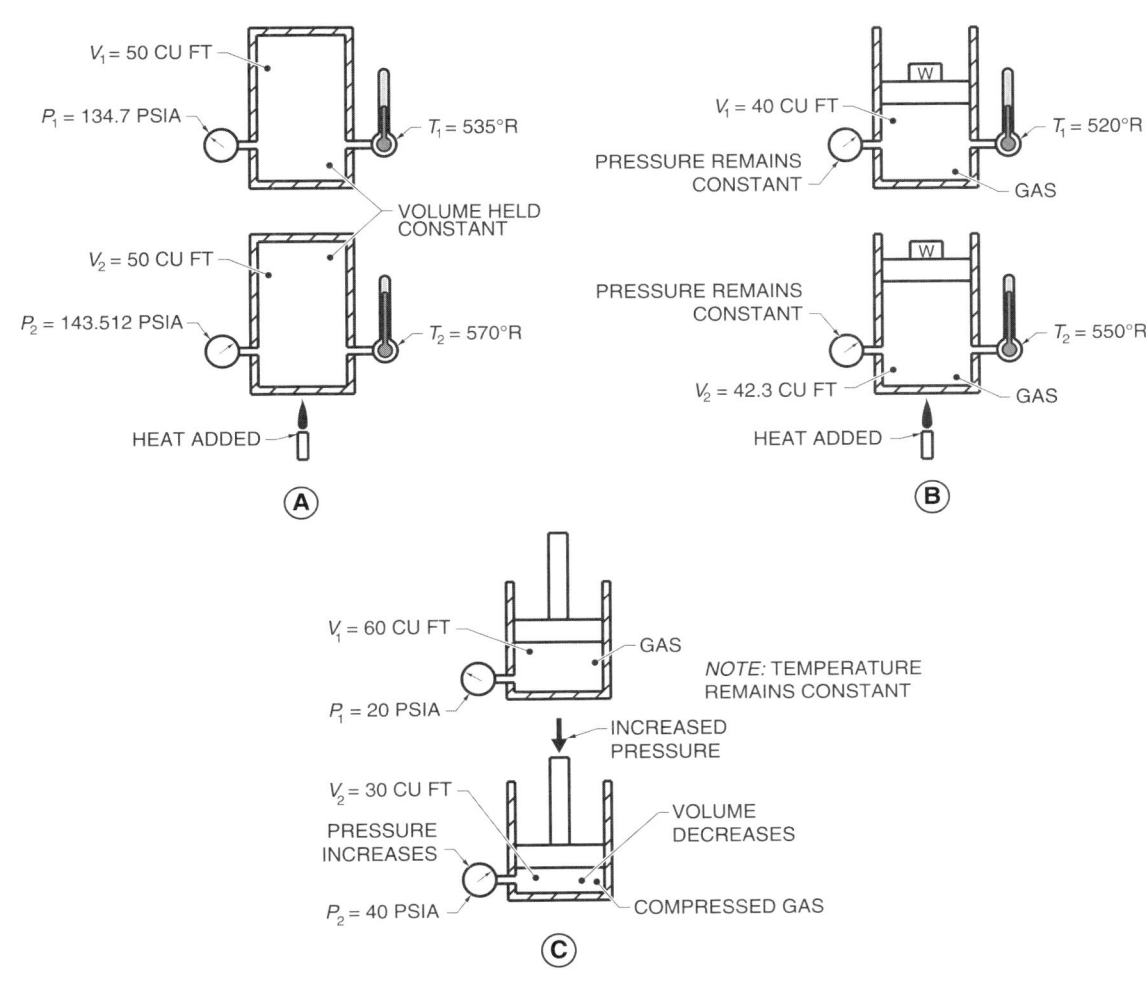

Desiccant Dryers

_____ 1. Check valves

_____ 2. Desiccant material

_____ 3. Reactivating dryer

_____ 4. Moist air inlet

_____ 5. Moist air outlet

_____ 6. Dryer operating

_____ 7. Purge valve

_____ 8. Dry air outlet

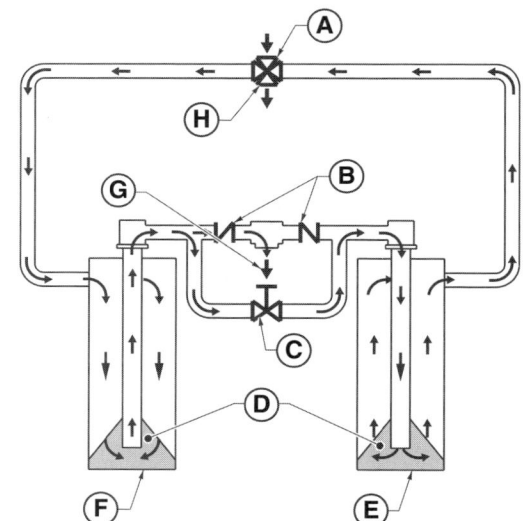

Fahrenheit/Rankine Temperatures

_____ 1. –460

_____ 2. 0

_____ 3. 32

_____ 4. 212

_____ 5. 492

_____ 6. 672

	°F	°R
WATER BOILS	A	B
WATER FREEZES	C	D
ABSOLUTE ZERO	E	F

Problems

_____ 1. Tank A has a volume of ___ cu ft.

_____ 2. Tank B has a capacity of ___ gal.

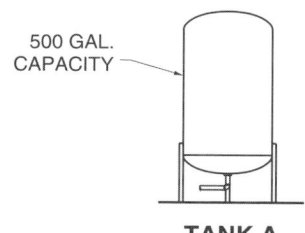

TANK A

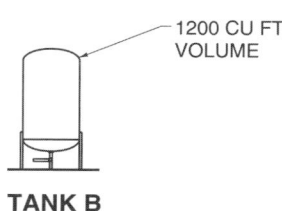

TANK B

3. The temperature on the Fahrenheit scale equals ___°R.

4. The final volume of a gas that occupies 120 cu ft at 60°F is ___ cu ft at 85°F.

5. The final pressure in a 100 cu ft tank holding a gas at 90 psig at 75°F is ___ psig when the temperature is increased to 112°F.

6. The final volume of 90 cu ft of air at 45 psia is ___ cu ft when expanded to 30 psia.

7. The ratio of compression is ___ in a compressor with an inlet pressure of 1.25 psi vacuum and a discharge pressure of 50 psig.

8. The final pressure at Tank C is ___ psia.

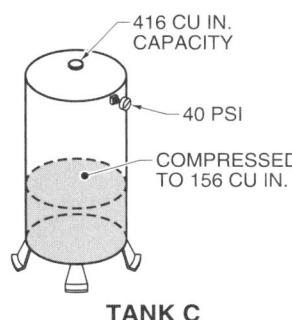

TANK C

9. Cylinder A has a capacity of ___ gal.

10. Fuel Can A has a volume of ___ cu ft.

11. Tub A has a volume of ___ cu ft.

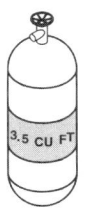 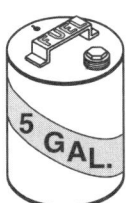 

CYLINDER A   **FUEL CAN A**   **TUB A**

_____ **12.** The absolute pressure within Tire A is ___ psia.

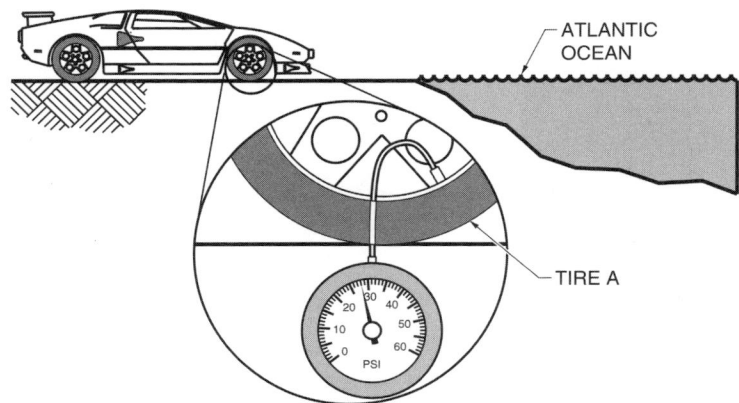

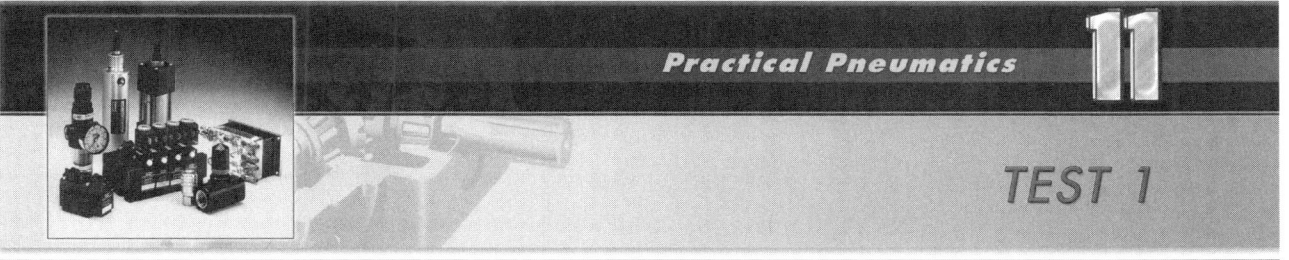

Practical Pneumatics

TEST 1

Name _____ Date _____

Industrial Mechanics

_____ 1. A pneumatic ___ transmits and controls energy through the use of a pressurized gas within an enclosed circuit.

_____ 2. A(n) ___ is a device that takes air from the atmosphere and compresses it to increase its pressure.

_____ 3. ___ pistons move forward and backward alternately.

_____ 4. A(n) ___ rod is the rod that connects the crankshaft to the piston.

_____ 5. A(n) ___ is the screw helix of a rotor.
　　　　　　　　　　　A. tongue
　　　　　　　　　　　B. ear
　　　　　　　　　　　C. leaf
　　　　　　　　　　　D. lobe

_____ 6. The main header of a pneumatic system should have a downward pitch of ___′ toward the drain pipe.
　　　　　　　　　　　A. 1″ per 1′
　　　　　　　　　　　B. 1″ per 10′
　　　　　　　　　　　C. 10″ per 10′
　　　　　　　　　　　D. none of the above

T　　F　　 7. A symbol is a graphic element which indicates a particular device, etc.

T　　F　　 8. Thread-sealing material should be placed in the female fitting only.

T　　F　　 9. The symbols for most of the components used in a pneumatic circuit are similar to those used in hydraulic circuits.

T　　F　　10. Logic is the science of correct reasoning.

_____ 11. A pressure ___ is a displacement control that alters displacement in response to pressure changes in a system.

_____ 12. A check valve allows flow in only ___ direction(s).

81

_____ 13. A binary system has ___ value(s).
    A. no
    B. one
    C. two
    D. any number of

_____ 14. A(n) ___ displacement compressor compresses a fixed quantity of air with each cycle.
    A. manual or automatic
    B. electric or gasoline
    C. vertical or horizontal tank
    D. none of the above

_____ 15. A(n) ___ compressor is a positive-displacement compressor that has multiple vanes located in an offset rotor.

_____ 16. A main ___ is the main air supply line that runs between the receiver and the circuits in a pneumatic system.

_____ 17. A(n) ___ is a device containing a porous substance through which a fluid can pass but particulate matter cannot.

_____ 18. Lubricators should be placed no more than ___′ from the lubricated components.

_____ 19. A pressure ___ is a value that restricts and/or blocks downstream air flow.

_____ 20. A(n) ___ is a device that converts electrical energy into a linear, mechanical force.

_____ 21. A(n) ___ is a device that senses a high- or low-pressure condition and relays an electrical signal to turn the compressor motor ON or OFF.
    A. pressure switch
    B. unloading valve
    C. safety relief valve
    D. none of the above

_____ 22. A(n) ___ is a device that senses a high-pressure condition and removes the compression energy.
    A. pressure switch
    B. unloading valve
    C. safety relief valve
    D. none of the above

_____ 23. A(n) ___ is a device that prevents excessive pressure from building up by venting air to the atmosphere.
    A. pressure switch
    B. unloading valve
    C. safety relief valve
    D. none of the above

T   F   **24.** An O-ring may be used as a static or a dynamic seal.

T   F   **25.** An air motor is an air-driven device that converts rotary mechanical energy into fluid energy.

T   F   **26.** A truth table lists the output condition of a logic element or combination of logic elements for every possible input condition.

T   F   **27.** Electric motors are less efficient than air motors.

T   F   **28.** Air motors are lighter than direct replacement electric motors.

T   F   **29.** The most popular air motor is the vane air motor.

T   F   **30.** Pneumatic circuits are generally cleaner than hydraulic circuits.

_____ **31.** A pneumatic ___ is a combination of air-operated components that are connected to perform work.

_____ **32.** A(n) ___ is the specific location of a spool within a valve which determines the direction of fluid flow through the valve.

_____ **33.** An air ___ is a device that converts compressed air energy into linear mechanical energy.

_____ **34.** A(n) ___ seal is a seal used as a gasket to seal nonmoving parts.

_____ **35.** A(n) ___ seal is a seal used between moving parts that prevents leakage or contamination.

_____ **36.** The prime mover in a helical screw compressor normally rotates the rotors at speeds between 3000 rpm and ___ rpm.
    A. 6000
    B. 8000
    C. 10,000
    D. 12,000

_____ **37.** A ¼″ hole in a pneumatic system that has an initial pressure of 80 psig loses about ___ scfm.
    A. 50.0
    B. 68.0
    C. 85.5
    D. 104

_____ **38.** A single-stage vane compressor normally has operating pressures between ___ psi and ___ psi.
    A. 0, 50
    B. 0, 100
    C. 0, 125
    D. 50, 125

84 INDUSTRIAL MECHANICS WORKBOOK

_____ 39. A two-stage vane compressor normally has operating pressures between ___ psi and ___ psi.
    A. 0, 50
    B. 0, 100
    C. 50, 100
    D. 50, 125

_____ 40. A hand pump used to pump air into a basketball or a bicycle tire is an example of a(n) ___ application.
    A. actuator
    B. check valve
    C. directional control valve
    D. gate valve

_____ 41. The outlet pressure produced by a 36″ D operating piston and a 24″ D ram operating at an inlet pressure of 350 psi is ___ psi.
    A. 155.55
    B. 225.00
    C. 525.00
    D. 787.50

Graphic Diagram — Pneumatic System

_____ 1. Electric motor
_____ 2. Filter
_____ 3. Aftercooler
_____ 4. Separator
_____ 5. Pressure switch
_____ 6. Compressor
_____ 7. Receiver
_____ 8. Safety relief valve
_____ 9. Manual shut-off valve

Graphic Diagram — Pneumatic Circuit

_____ 1. Flow from compressor

_____ 2. Actuator

_____ 3. Regulator

_____ 4. Lubricator

_____ 5. Filter

_____ 6. Directional control valve

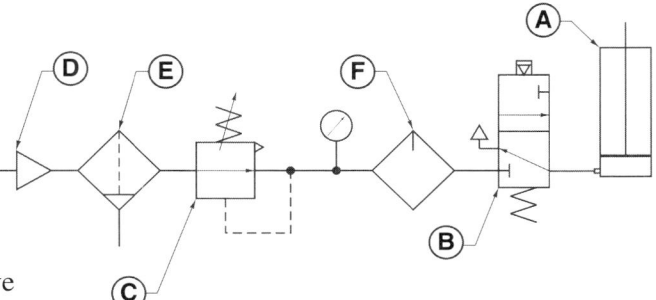

Safety Relief Valves

_____ 1. Spring

_____ 2. System pressure

_____ 3. Valve vent port

_____ 4. Seat

_____ 5. Poppet

_____ 6. Pull ring

_____ 7. Vent to atmosphere

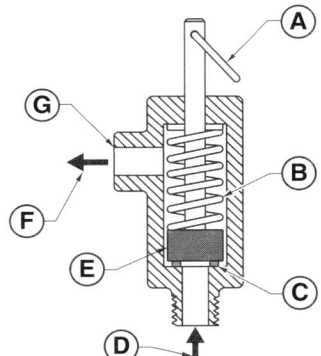

Pneumatic System Symbols

_____ 1. Pipe slope in direction of flow

_____ 2. Liquid separator with automatic drain

_____ 3. Filter with manual drain

_____ 4. Lubricator with manual drain

_____ 5. Gate valve

_____ 6. Pressure gauge

_____ 7. Quick disconnect

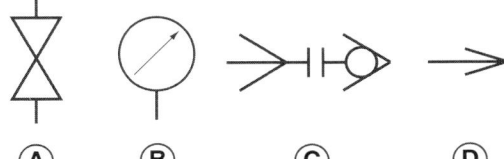

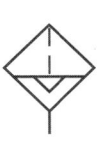
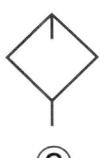

86 INDUSTRIAL MECHANICS WORKBOOK

Problems

_____ 1. The outlet pressure produced by the intensifier is ___ psi.

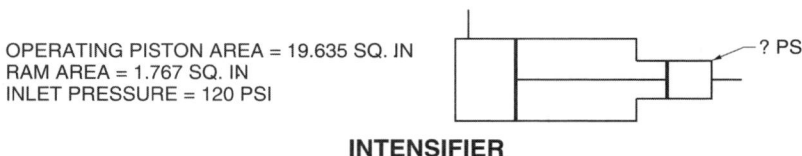

OPERATING PISTON AREA = 19.635 SQ. IN
RAM AREA = 1.767 SQ. IN
INLET PRESSURE = 120 PSI

INTENSIFIER

_____ 2. The main header should drop ___″ from the receiver to the moisture drop pipe.

_____ 3. The pressure drop in a pneumatic system with 119′ of 2″ Schedule 40 pipe, one 2″ globe valve, two 2″ 45° elbows, and one 25 μ filter that has a working pressure of 110 psi and an airflow rate of 70 scfm is ___ psi.

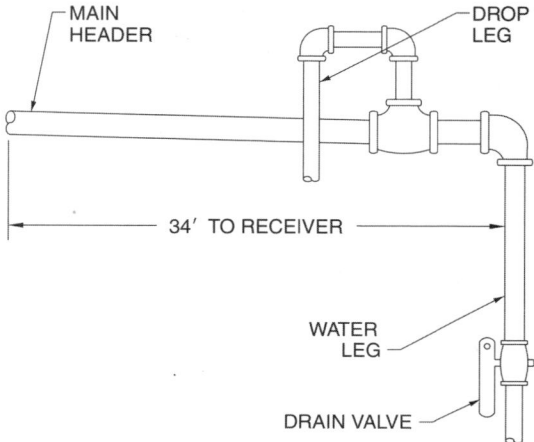

Lubrication 12

TEST 1

Name _____ Date _____

Industrial Mechanics

_____ 1. ___ is the process of maintaining a fluid film between solid surfaces to prevent their physical contact.

_____ 2. The ___ of friction is the measure of frictional force between two surfaces in contact.

_____ 3. ___ is a chemical adsorption process in which weak chemical bonds are formed between liquid or gas molecules and solid surfaces.

_____ 4. ___ gases are gases that lack active properties.

T F 5. Walking requires friction between the feet and floor in order to move.

T F 6. Lubrication generally involves coating surfaces with a material that has a higher coefficient of friction than the original surfaces.

T F 7. Friction occurs when an object in contact with another object tries to move.

_____ 8. Lubricants are used to ___.
  A. reduce friction
  B. prevent wear
  C. prevent corrosion
  D. all of the above

_____ 9. Liquid lubricants include ___.
  A. animal/vegetable oils
  B. petroleum fluids
  C. synthetic fluids
  D. all of the above

T F 10. Animal and vegetable oils are used mostly in the food industry.

T F 11. Petroleum is composed of 12% carbon and 85% hydrogen, with a small amount of other elements.

T F 12. Animal and vegetable oils contain fatty acids.

_____ 13. Earth's plants release approximately ___ million tons of hydrocarbons into the air each year.

_____ 14. ___ is the measurement of the resistance of a fluid's molecules to move past each other.

_____ 15. Shear ___ is a liquid's ability to remain as a separator between solids in motion.

_____ 16. A(n) ___ is a device that vaporizes elements in the oil sample into light.

T F 17. The grease used in a centralized system should be one grade softer than is otherwise required.

T F 18. Sealed bearings should be relubricated on a regularly-scheduled basis.

T F 19. Shear stress is stress in which the material on one side of a surface pushes on the material on the other side of the surface with a force perpendicular to the surface.

T F 20. Synthetic lubricants are generally higher priced than petroleum lubricants.

_____ 21. During startup of a machine, oil ___.
    A. is cool
    B. does not flow easily
    C. A and B
    D. none of the above

_____ 22. ___ action is the action by which the surface of a liquid is elevated on a material due to its relative molecular attraction.
    A. Submission
    B. Polymeric
    C. Capillary
    D. none of the above

T F 23. Lubrication contamination is the main cause of mechanical system failure.

T F 24. Oil that is contaminated with water has a clear appearance.

T F 25. Petroleum is formed by an evolutionary process that takes many millions of years.

Oil Application Systems

_____ 1. Submersion–Splash

_____ 2. Submersion–Chain

_____ 3. Submersion–Ring

_____ 4. Drip

_____ 5. Wick

_____ 6. Centralized

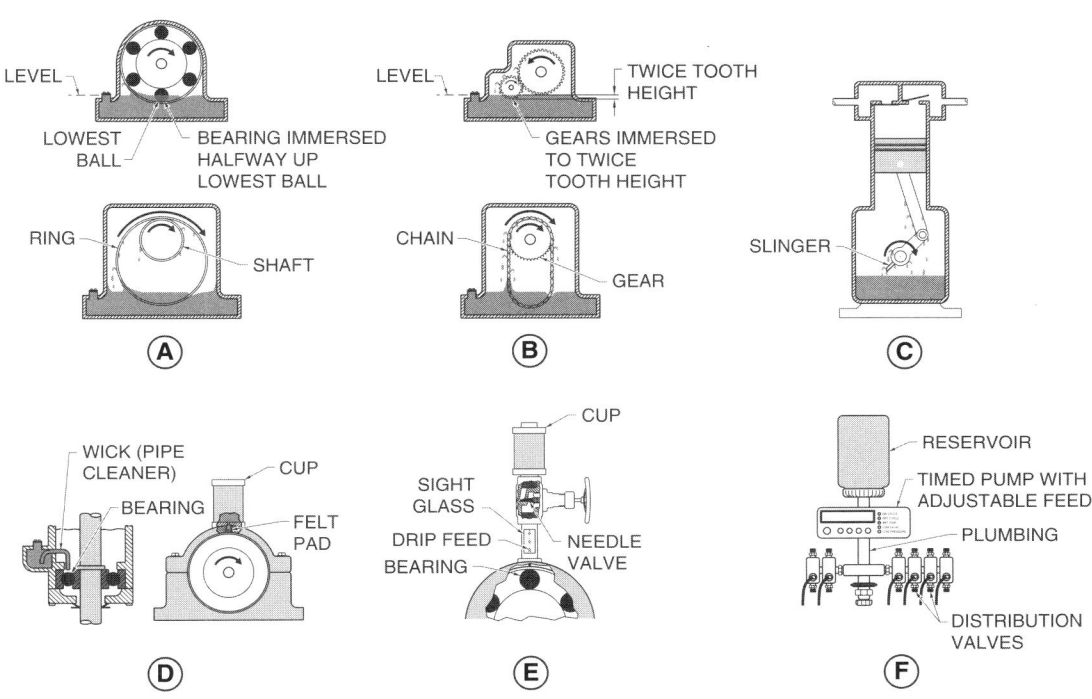

Grease Thickeners

_____ 1. Aluminum soap

_____ 2. Calcium soap

_____ 3. Lithium soap

_____ 4. Clay

_____ 5. Fiber

A. used for extreme temperatures

B. added to resist being thrown off

C. offers clarity

D. is water-resistant

E. allows high temperature use

Lubricant Additives

_____	1. Oxidation inhibitors	A. prevent rust
_____	2. Rust inhibitors	B. improve film strength
_____	3. Fatty materials	C. prevent galling
_____	4. Powdered lead or graphite	D. separate out water
_____	5. Viscosity index improvers	E. provide long bearing or gear life
_____	6. Demulsifiers	F. ease machine movement in cold weather

Oil Groups/Application

_____	1. Group A	A. machine tools
_____	2. Group B	B. automotive
_____	3. Group C	C. reciprocating engines
_____	4. Group D	D. turbojet engines
_____	5. Group E	E. gear trains and transmissions
_____	6. Group F	F. marine propulsions and stationary power turbines

Problems

_____ 1. The coefficient of friction of Object A is ___.

[Diagram: 10 LB force applied to OBJECT A weighing 30 LB]

_____ 2. A 40 lb force is required to overcome the frictional force between a 75 lb object and the surface upon which it is resting. The coefficient of friction is ___.

T   F   3. The coefficient of friction of Object B is 5.

[Diagram: 8 LB force applied to OBJECT B weighing 40 LB]

Lubrication 12
TEST 2

Industrial Mechanics

T F **1.** Greater force is required to move a body from a static condition than is required to keep it in a kinetic condition.

_____ **2.** ___ lubrication is the condition of lubrication in which the friction between two surfaces in motion is determined by the properties of the surfaces and the properties of the lubricant other than viscosity.
  A. Area
  B. Material
  C. Surface
  D. none of the above

_____ **3.** A(n) ___ lubricant is a lubricant that uses pressurized air to separate two surfaces.
  A. chemical
  B. metal
  C. gas
  D. none of the above

_____ **4.** A petroleum fluid is a fluid consisting of ___.

T F **5.** The flow rate is the most important property of a lubricant.

T F **6.** Under basic conditions, as the temperature of oils increases, their viscosity also increases.

T F **7.** Lubricating oil is given an SAE viscosity rating based on its ability to flow at a specific temperature.

T F **8.** A 10 weight oil is thicker than a 40 weight oil.

_____ **9.** A ___ solid is a solid that is finely ground in order to be spread.
  A. disposed
  B. dispersed
  C. dispelled
  D. displaced

92 INDUSTRIAL MECHANICS WORKBOOK

_____ 10. A ___ is the result of a chemical reaction in which two or more small molecules combine to form larger molecules.
    A. polygon
    B. polymer
    C. either A or B
    D. none of the above

T   F   11. Solid lubricants such as graphite shear easily between sliding surfaces.

T   F   12. All greases exhibit a dropping point.

T   F   13. Water that mixes with lubricants increases the effectiveness of the lubricant.

_____ 14. Wear particle ___ is the study of wear particles present in lubricating oil.

_____ 15. Fluid lubricants must create a(n) ___ between material surfaces to prevent contact with each other.

T   F   16. As temperatures increase, greases become softer.

T   F   17. Graphite has high shearing forces.

T   F   18. Gas lubricants are commonly used in low-friction, high-speed applications.

_____ 19. Approximately ___% of all lubricants used today are petroleum based.

_____ 20. Oil film thickness ___ with an increase in oil temperature.

Petroleum

_____ 1. Pilot oil well

_____ 2. Crude oil pumped from well

_____ 3. Soil

_____ 4. Porous rock

_____ 5. Nonporous rock

_____ 6. Bedrock

_____ 7. Crude oil

_____ 8. Natural gas

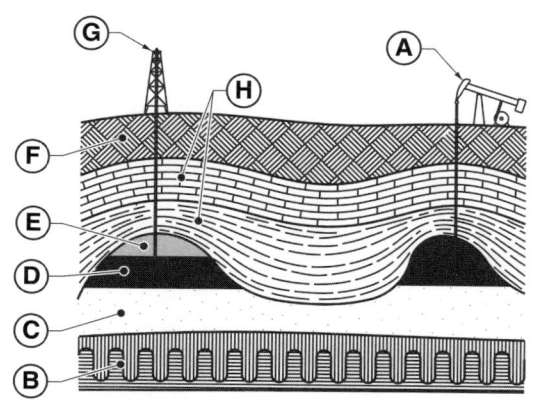

Motor Regreasing

_____ 1. Wipe grease fitting, drain plug, and grease gun nozzle.

_____ 2. Remove drain plug and clean.

_____ 3. Add grease until grease is expelled from drain plug port.

_____ 4. Run motor to expel excess grease.

_____ 5. Clean and replace drain plug.

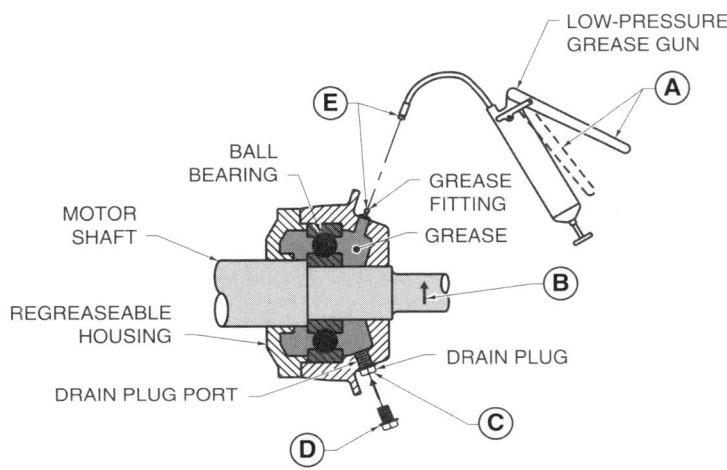

Grease Application Methods

_____ 1. Grease cup

_____ 2. Grease gun

_____ 3. Centralized system

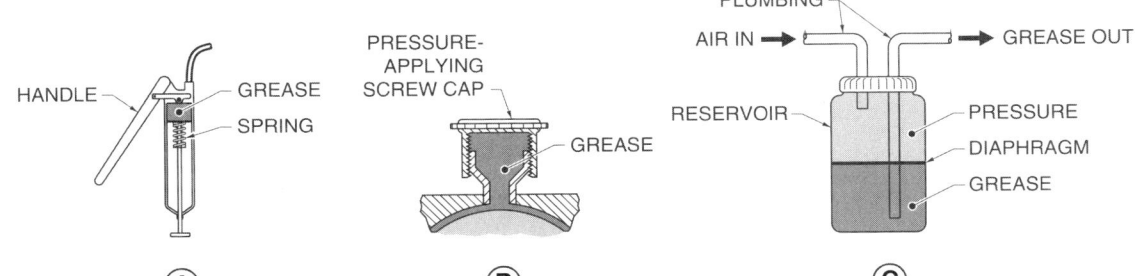

NLGI Grease Grades

T F   **1.** The higher the NLGI number, the stiffer the grease.

T F   **2.** The higher the NLGI number, the more penetration it has.

T F   **3.** Grade 1 is softer than Grade 00.

_____   **4.** Grade 5 will penetrate approximately one-half as much as Grade ___.

T F   **5.** Grade 000 has three times the penetration of Grade 0.

T F   **6.** Grades 0, 1, and 2 are the most widely used in industry.

T F   **7.** For maximum penetration, a higher NLGI grade of grease should be used.

_____   **8.** NLGI Grade ___ has a penetration range from .68″ to .80″.

NLGI GRADE GRADES		
NLGI Grade	Penetration*	Stiffness
000	1.75–1.87	VERY SOFT ↕ VERY HARD
00	1.57–1.69	
0	1.32–2.30	
1	1.22–1.33	
2	1.04–1.16	
3	.86–.98	
4	.68–.80	
5	.51–.62	
6	.33–.45	

* in in.

Coefficients of Friction

T   F   1. Greater force is required to move a body from rest than is required to keep it in motion.

T   F   2. The static condition relating to coefficient of friction refers to the forces required to start a solid object in motion.

_____ 3. The coefficient of friction required to start the movement of a piece of copper resting on an unlubricated copper plate is ___.

_____ 4. The coefficient of friction required to maintain the movement of a piece of copper on an unlubricated copper plate is ___.

_____ 5. A steel object resting on an unlubricated piece of steel weighs 10 lb. A force of ___ lb is required to start it in motion.

_____ 6. The coefficient of friction of a 500 lb object resting on a horizontal surface that requires 125 lb of force to move is ___.
    A. 0.03
    B. 0.25
    C. 0.40
    D. 4.0

_____ 7. The coefficient of friction of a 3500 lb object resting on a horizontal surface that requires 225 lb of force to move is ___.
    A. 0.06
    B. 0.16
    C. 0.64
    D. 15.6

_____ 8. The coefficient of friction of a 2200 lb object resting on a horizontal surface that requires 931 lb of force to move is ___.
    A. 0.42
    B. 0.84
    C. 0.95
    D. 1.1

_____ 9. The coefficient of friction of a 51 lb object resting on a horizontal surface that requires 6 lb of force to move is ___.
    A. 0.01
    B. 0.12
    C. 0.85
    D. 8.5

_____ **10.** The coefficient of friction of a 23¼ lb object resting on a horizontal surface that requires 2.11 lb of force to move is ___.
   A. 0.09
   B. 0.11
   C. 0.92
   D. 11

T   F   **11.** Gas lubricants can operate in temperatures from –400°F to over 3500°F.

T   F   **12.** The coefficient of friction for unlubricated copper-to-copper surfaces is higher than that for unlubricated steel-to-steel surfaces.

COEFFICIENTS OF FRICTION

Material	Unlubricated		Lubricated*	
	Static	Kinetic	Static	Kinetic
Steel-to-Steel	.8	.4	.16	.02
Copper-to-Copper	1.5	.3	.08	.02
Aluminum-to-Aluminum	1.3	—	.3	—
Nylon-to-Nylon	.3	.1	—	—
Teflon-to-Teflon	.04	.03	—	—
Graphite-to-Graphite	.1	.06	—	—

*values are approximations and vary according to lubricant type

Bearings 13
TEST 1

Name _____ Date _____

Industrial Mechanics

_____ 1. A(n) ___ is a machine part that supports another part, such as a shaft, which rotates or slides in or on it.

_____ 2. ___ life is the maximum useful life of a bearing.

_____ 3. A(n) ___ load is a load in which the applied force is parallel to the axis of rotation.

_____ 4. ___ life is the length of service received from a bearing.

_____ 5. The ___ is the track on which the balls of a bearing move.
   A. cup
   B. cone
   C. race
   D. none of the above

_____ 6. Under normal load conditions, ball bearings generally have ___″ interference per inch of shaft when the inner race is press fit.
   A. .00025
   B. .0025
   C. .025
   D. .25

T   F   7. Doubling the load on a bearing increases its service life by 6 to 8 times.

T   F   8. A better finish on a bearing produces less friction.

T   F   9. Ball bearings are installed with one ring being a press fit and the other ring a push fit.

T   F   10. Needle bearings are generally press fit.

_____ 11. A(n) ___ bearing is a bearing in which the shaft turns and is lubricated by a sleeve.

_____ 12. ___ is the flaking away of metal pieces due to metal fatigue.

_____ 13. The ___ of grease is the temperature at which the oil in grease separates from the thickener and runs out, leaving just the thickener.

_____ 14. Bearing surfaces that are ___ appear as worn surfaces on one side or opposing sides of a bearing.

_____ 15. ___ is a bonding, shearing, and tearing away of material from two contacting, sliding metals.

_____ 16. ___ damage is bearing damage due to axial force.

_____ 17. ___ is the elongated and rounded grooves or tracks left by the etching of each roller on the rings of an improperly grounded roller during welding.

_____ 18. Precision class bearings are generally marked with their high points of ___.

_____ 19. ___ play is the total amount of axial movement of a shaft.

_____ 20. A ___-contact bearing is a bearing composed of rolling elements between an inner and outer ring.

T   F   21. Plain bearings may support radial and axial loads.

T   F   22. Bearing installation is generally more difficult than bearing removal.

T   F   23. Bearings should never be struck with a hammer.

T   F   24. Solid or caked lubricant is a sign that bearings have overheated.

T   F   25. As the temperature of steel increases, it discolors, turning from silver to blue to black.

T   F   26. Prelubricated bearings may be heated for installation.

T   F   27. Never apply pressure on the outer ring if the inner ring is press fit and never apply pressure on the inner ring if the outer ring is press fit.

_____ 28. A threaded cup ___ is a tapered bearing gap adjusting device that is used to adjust shaft endplay by controlling the amount of clearance between tapered roller bearings.

_____ 29. A(n) ___ bore bearing is a bearing whose bore varies in diameter from the face to the back of the bearing.

T   F   30. Roller-contact bearings include ball, roller, and needle bearings.

T   F   31. Needle bearings are designed primarily for relatively low radial loads.

T   F   32. Babbitt metals are the best metals for plain bearing loads.

T   F   33. False Brinell damage is bearing damage caused by forces passing from one ring to the other through the balls or rollers.

T   F   34. A machine should never be grounded by connecting a wire from the machine to a gas or oil pipe.

_____ 35. A(n) ___ is the part of a shaft, such as an axle or spindle, that moves in a plain bearing.

36. A(n) ___ bearing is a rolling-contact bearing in which the load is transmitted perpendicularly to the axis of shaft rotation.
    A. angular contact
    B. Conrad
    C. loading
    D. radial

37. A(n) ___ slot is a groove or notch on the inside wall of each bearing ring to allow insertion of balls.
    A. angular
    B. ball
    C. loading
    D. ring

38. Bearings that are designed for ___ loads must be installed in only one direction to prevent the load from separating the bearing components.
    A. compression
    B. low-weight
    C. tensile
    D. thrust

39. Double-row bearings are also known as ___ bearings.
    A. ball
    B. duplex
    C. needle
    D. plain

40. Cylindrical roller bearings are used in high-speed, high-load applications and may contain as many as ___ rows of rollers.
    A. four
    B. six
    C. eight
    D. twelve

41. ___ bearings are suited for chemical or high-temperature applications and require no lubrication.
    A. Antimony
    B. Bronze
    C. Copper-lead
    D. Nylon and Teflon®

42. ___ bearings are designed to operate in temperatures exceeding 700°F and withstand 300 psi of load force without a lubricant.
    A. Aluminum
    B. Carbon-graphite
    C. Lead
    D. Nylon and Teflon®

_____ **43.** ___ damage in ball bearings appears as marks on the shoulder or upper portion of the inner and outer races or as heavy wear at the bearing ends of plain bearings.
A. Misalignment
B. Pitting
C. Spalling
D. Thrust

_____ **44.** When mounting bearings through the use of controlled temperatures, temperatures over ___°F may reduce the hardness of bearing metals, resulting in early failure.
A. 250
B. 300
C. 350
D. 400

_____ **45.** Bearings are classified as ___ or plain bearings.
A. friction
B. direct
C. rolling-contact
D. static

Bearing Failure

_____ 1. Dark, discolored metals indicate ___.   **A.** improper fit or assembly

_____ 2. Rusting surfaces indicate ___.   **B.** high temperatures

_____ 3. Split or fractured rings indicate ___.   **C.** high moisture and/or improper lubrication

Angular Contact Bearing Use

_____ 1. Face-to-face

_____ 2. Back-to-back

_____ 3. Separated face-to-face

Bearing Loads

_____ 1. Radial load

_____ 2. Axial load

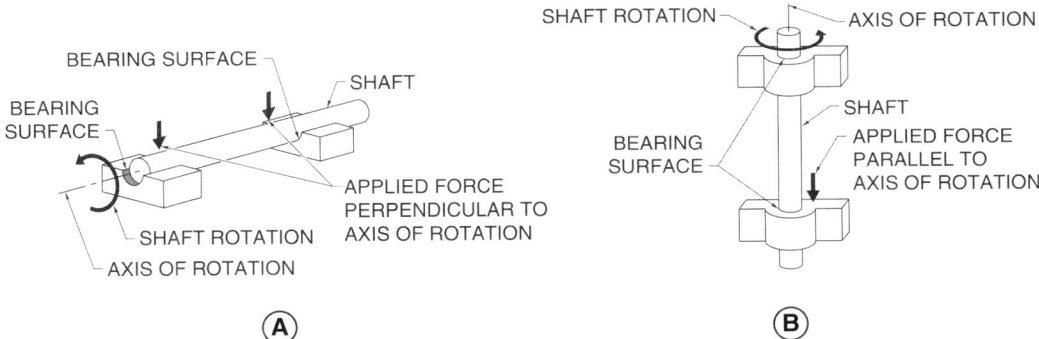

Rolling-Contact Bearings

_____ 1. Ball

_____ 2. Needle

_____ 3. Roller

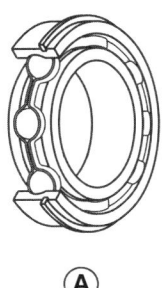 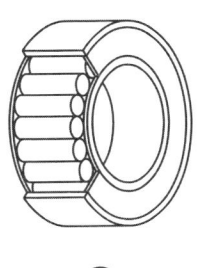 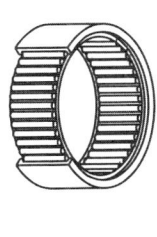

Ball Bearings

_____ 1. Single-row radial

_____ 2. Single-row angular-contact

_____ 3. Double-row radial or axial

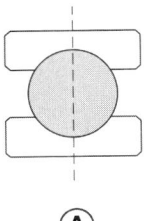 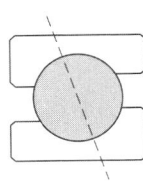 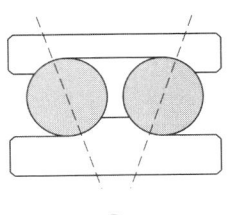

Flexible Belt Drives 14

TEST 1

Name _____ Date _____

Industrial Mechanics

_____ 1. A(n) ___ is an endless power transmission belt with a trapezoidal cross section.

_____ 2. ___ V-belts are designated as A, B, C, D, or E.

_____ 3. ___ V-belts are designated as 3V, 5V, or 8V.

_____ 4. V-belts run in a(n) ___ (sheave) with a V-shaped groove.

_____ 5. Angular misalignment of a pulley must not exceed ___°.

T   F   6. Too little tension on a belt can cause the belt to slip.

_____ 7. ___ is the process of preventing the flow of energy from a power source to a piece of equipment.

_____ 8. ___ is the process of placing a tag on a power source that warns others not to restore energy.

_____ 9. ___ is the process of placing a solid object in the path of a power source to prevent accidental energy flow.

T   F   10. For optimum efficiency, a V-belt should touch the bottom of the pulley.

T   F   11. V-belt forces remain constant as the belt bends around the pulley.

T   F   12. A fixed bore pulley is a machine-bored one-piece pulley.

T   F   13. Pulleys should be placed as close as possible to the shaft bearing to prevent overhung loads.

_____ 14. ___ misalignment is a condition where two shafts are parallel but the pulleys are not on the same axis.
   A. Offset
   B. Non-parallel
   C. Angular
   D. none of the above

103

_____ 15. A ___ V-belt is a belt designed to transmit power from the top and bottom of the belt.
  A. ½
  B. top/bottom
  C. single
  D. double

_____ 16. A(n) ___ belt drive is a mechanism that transmits motion from one shaft to another and allows the speed of the shafts to be varied.

_____ 17. A(n) ___ groove gauge is a gauge that has a male form to determine the size of a pulley and a female form to determine the size of a belt.

_____ 18. A(n) ___ belt is a belt designed for positive transmission and synchronization between the drive shaft and the driven shaft.

_____ 19. Belt ___ length is the total length of the timing belt measured at the belt pitch line.

_____ 20. A(n) ___ value is a designated or theoretical value that may vary from the actual value.

_____ 21. The recommended torque value for a size W hub is ___ lb/ft.
  A. 200
  B. 300
  C. 400
  D. 500

_____ 22. The recommended torque value for a size SH hub is ___ lb/ft.
  A. 6
  B. 10
  C. 50
  D. 100

_____ 23. Offset alignment must be within ___″ per foot of drive center distance.
  A. 1/16
  B. 1/10
  C. 1/8
  D. 3/16

_____ 24. The proper belt deflection of an assembly using a 6″ pulley and a 3″ pulley having a span length of 12″ is ___″.
  A. 0.046
  B. 0.094
  C. 0.140
  D. 0.187

25. The proper belt deflection of an assembly using a 15″ pulley and a 6″ pulley having a span length of 42″ is ___″.
    A. 0.234
    B. 0.327
    C. 0.655
    D. 0.765

26. The proper belt deflection of an assembly using a 36″ pulley and a 12″ pulley having a span length of 24″ is ___″.
    A. 0.749
    B. 0.562
    C. 0.374
    D. 0.187

27. A trapezoidal timing belt with a double extra heavy cross section is classified as ___.
    A. CC
    B. DD
    C. XXL
    D. XXH

28. A trapezoidal timing belt with a mini extra light cross section is classified as ___.
    A. AA
    B. BB
    C. MXL
    D. MXH

29. A trapezoidal timing belt with a heavy cross section is classified as ___.
    A. BB
    B. CC
    C. H
    D. XH

30. The speed of a 15″ pulley driven by a 36″ drive pulley at 60 rpm is ___ rpm.
    A. 9
    B. 25
    C. 60
    D. 144

31. The speed of a 24″ pulley driven by a 48″ drive pulley at 175 rpm is ___ rpm.
    A. 87.5
    B. 175
    C. 350
    D. 525

106 INDUSTRIAL MECHANICS WORKBOOK

_____ 32. The speed of a 6″ pulley driven by an 18″ drive pulley at 100 rpm is ___ rpm.
   A. 100
   B. 300
   C. 333
   D. 400

Variable-Speed Belt Drives

_____ 1. The variable-speed belt drive is at ___ speed.

_____ 2. Spring

_____ 3. Shaft

_____ 4. V-belt

_____ 5. Pitch diameter

_____ 6. Central sheave

_____ 7. Cone-faced pulley flanges

_____ 8. Set screw

Timing Belt Tooth Profiles

_____ 1. Trapezoidal

_____ 2. Double trapezoidal

_____ 3. Curvilinear

_____ 4. Modified curvilinear

Recommended Minimum Pulley Diameters

_____ 1. The recommended minimum pulley diameter for a 7½ HP motor running at 1160 rpm is ___″.

_____ 2. The recommended minimum pulley diameter for a 15 HP motor running at 1750 rpm is ___″.

_____ 3. The recommended minimum pulley diameter for a 1 HP motor running at 1750 rpm is ___″.

_____ 4. The recommended minimum pulley diameter for a 100 HP motor running at 870 rpm is ___″.

Motor HP	Motor Speed**			
	870	1160	1750	3500
½	2.2	—	—	—
¾	2.4	2.2	—	—
1	2.4	2.4	2.2	—
1½	2.4	2.4	2.4	2.2
2	3.0	2.4	2.4	2.4
3	3.0	3.0	2.4	2.4
5	3.8	3.0	3.0	2.4
7½	4.4	3.8	3.0	3.0
10	4.4	4.4	3.8	3.0
15	5.2	4.4	4.4	3.8
30	6.8	6.8	5.2	—
75	10.0	10.0	8.6	—
100	12.0	10.0	8.6	—

RECOMMENDED MINUMUM PULLEY DIAMETERS*

*in in.
** in rpm

Pulley Misalignment

_____ 1. Angular

_____ 2. Offset

_____ 3. Nonparallel

Problems

_____ 1. The belt length for two pulleys 6″ and 10″ in diameter that are 42″ apart at their centers is ___″.

_____ 2. The belt length required at A is ___″.

_____ 3. The proper belt deflection at B is ___″.

_____ 4. The driven pulley speed at C is ___ rpm.

_____ 5. The drive pulley speed at D is ___ rpm.

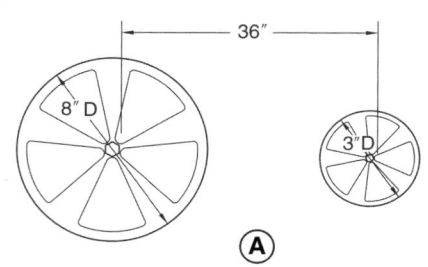

A

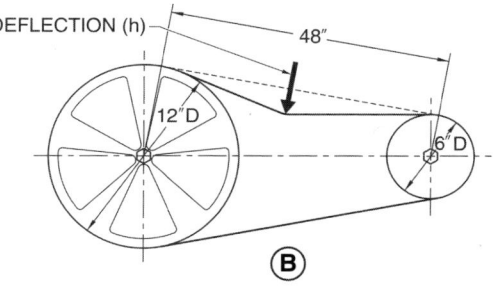

B

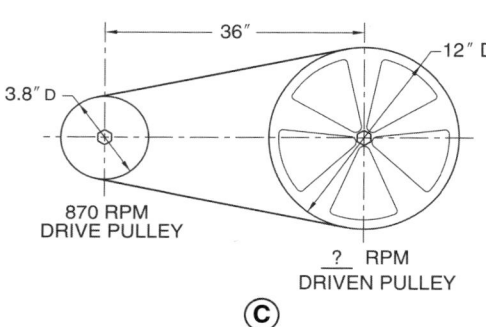

C

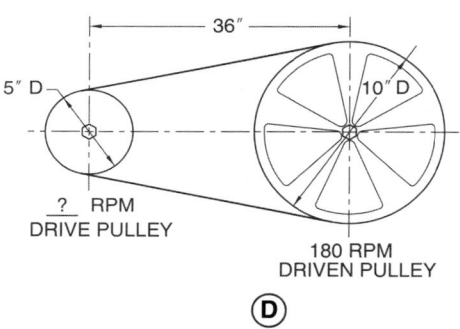

D

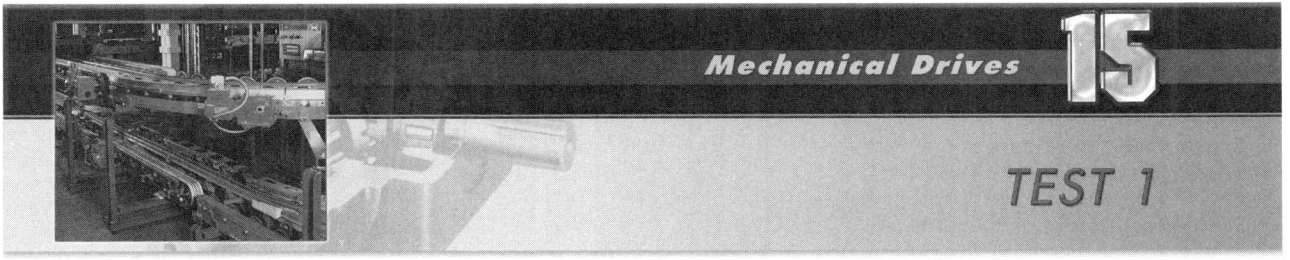

Mechanical Drives — TEST 1

Name _____ Date _____

Industrial Mechanics

_____ 1. A(n) ___ is a toothed machine element used to transmit motion between rotating shafts.

_____ 2. A(n) ___ drive is a system by which power is transmitted from one point to another.

_____ 3. ___ is the twisting force of a shaft.
   A. Rotation
   B. Torque
   C. Shearing
   D. Bending

_____ 4. To find lb-in of torque when lb-ft of torque is known, ___.
   A. add 12
   B. subtract 12
   C. multiply by 12
   D. divide by 12

_____ 5. Horsepower is a unit of power equal to ___.
   A. 550 lb-ft/sec
   B. 746 W
   C. 33,000 lb-ft/min
   D. all of the above

_____ 6. A(n) ___ gear is any gear that turns or drives another gear.

_____ 7. A(n) ___ is the relationship between two quantities of terms.

_____ 8. A(n) ___ gear is a gear that transfers motion and direction in a gear train, but does not change speeds.

T   F   9. The colon is the symbol used to indicate a relation between terms.

T   F   10. Adding an idler gear between a driven and drive gear changes the direction of rotation of the driven gear.

T F **11.** The tooth form of a rack gear consists of two flat surfaces.

T F **12.** Backlash is the play between mating gear teeth.

_____ **13.** A(n) ___ gear is a gear that has straight teeth that are parallel to the shaft axis.

_____ **14.** Helical gear drive angles may be anywhere from 0° to ___°.
  A. 30
  B. 60
  C. 90
  D. 120

_____ **15.** A compound gear train is ___ or more sets of gears where ___ gear(s) is/are keyed and rotate(s) on one common shaft.
  A. one; one
  B. one; two
  C. two; one
  D. two; two

_____ **16.** A tooth ___ is the shape or geometric form of a tooth in a gear when seen as its side profile.

_____ **17.** ___ pitch is the ratio of the number of teeth in a gear to the diameter of the gear's pitch circle.

_____ **18.** ___ depth is the depth of engagement of two gears.

_____ **19.** A(n) ___ gear is a gear that connects shafts at an angle in the same plane.

_____ **20.** ___ is the action or process of eating or wearing away gradually by chemical action.

_____ **21.** A(n) ___ fracture is a breaking or tearing of gear teeth.

T F **22.** Rack teeth are gear teeth used to produce linear motion.

T F **23.** Spur gears are quieter and smoother running than helical gears.

T F **24.** Under normal conditions, the maximum operating temperature of a gear drive should not exceed 211°F.

T F **25.** Gear manufacturers design certain parts of a gear train to wear out or break sooner than others.

Gear Terminology

_____ 1. Center distance

_____ 2. Pinion

_____ 3. Gear

_____ 4. Pitch circle

_____ 5. Outside diameter

_____ 6. Base diameter

_____ 7. Base circle

_____ 8. Circular pitch

_____ 9. Working depth

_____ 10. Line of action

_____ 11. Clearance

_____ 12. Tooth profile (Involute)

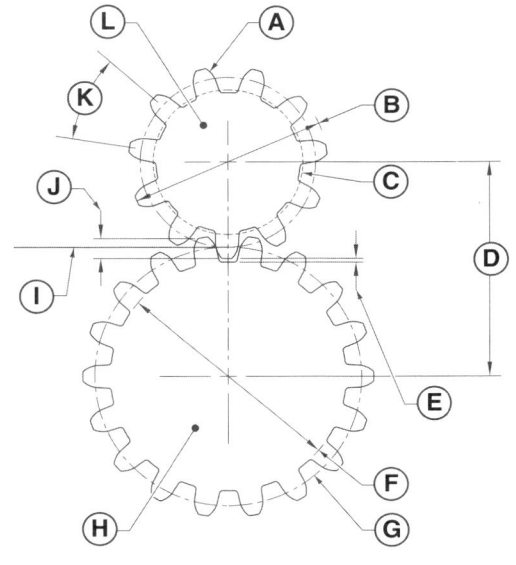

Gears

_____ 1. Miter

_____ 2. Worm

_____ 3. Bevel

_____ 4. Hypoid

_____ 5. Helical

_____ 6. Spur

_____ 7. Rack and pinion

_____ 8. Herringbone

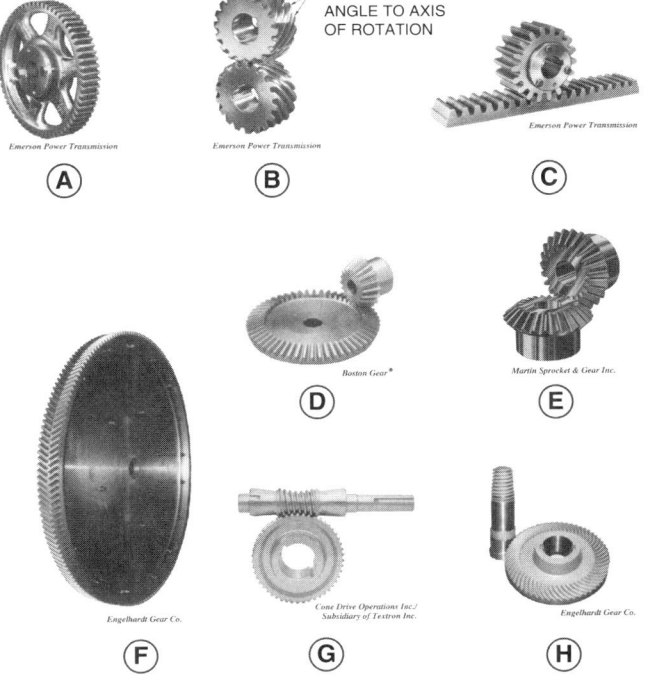

Gear Wear

_____ 1. Abrasive wear

_____ 2. Corrosive wear

_____ 3. Electrical pitting

_____ 4. Fatigue wear

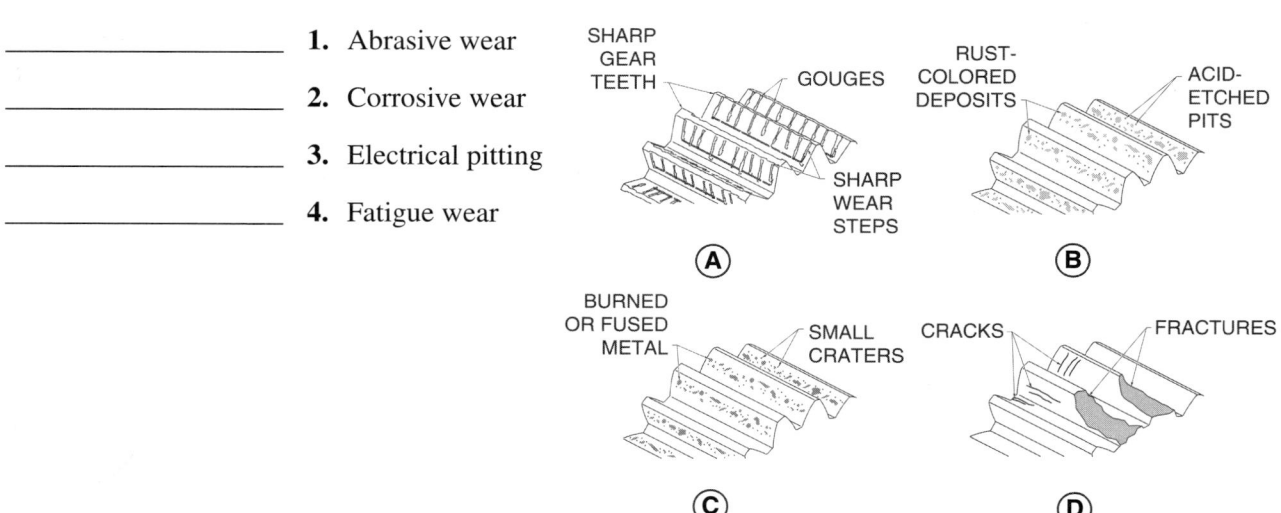

Problems

_____ 1. A torque of ___ lb-ft is developed when a 75 lb force is applied at the end of a 3′ lever arm.

_____ 2. The available torque supplied by a 1.5 HP, 1750 rpm motor is ___ lb-ft.

_____ 3. ___ HP is required to turn Winch A.

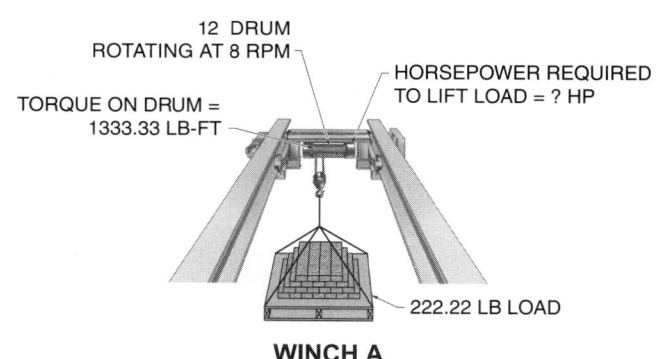

WINCH A

_____ 4. The speed of a 60 tooth driven gear is ___ rpm when the drive gear has 20 teeth and rotates at 120 rpm.

_____ 5. Gear A is rotating in a(n) ___ direction.

_____ 6. Gear A is rotating at ___ rpm.

_____ 7. A driven gear rotating at 36 rpm requires ___ teeth if the 48 tooth drive gear rotates at 24 rpm.

_____ 8. The diametral pitch (DP) of Gear C is ___.

DRIVEN GEAR A ROTATING AT _?_ RPM — 50 TEETH
18 TEETH
DRIVE GEAR B ROTATING AT 100 RPM IN COUNTERCLOCKWISE DIRECTION

GEARS A AND B

GEAR C 50 TEETH
PITCH CIRCLE = 4″

GEAR C

_____ 9. The available torque supplied by a ¾ HP, 3450 rpm blower motor is ___ lb-ft.

_____ 10. The available torque supplied by a ⅓ HP, 1800 rpm DC motor is ___ lb-ft.

_____ 11. The available torque supplied by a ½ HP, 1725 rpm magnetic brake motor is ___ lb-ft.

_____ 12. The horsepower required to turn a 48″ drum at 72 rpm with a 3200 lb load on the winch is ___ HP.

_____ 13. The horsepower required to turn a 36″ drum at 6 rpm with a 220 lb load on the winch is ___ HP.

_____ 14. The horsepower required to turn a 24″ drum at 24 rpm with a 51 lb load on the winch is ___ HP.

_____ 15. The speed of an 80 tooth gear driven by a 12 tooth drive gear that operates at 60 rpm is ___ rpm.

_____ **16.** The speed of a 100 tooth gear driven by a 32 tooth drive gear that operates at 120 rpm is ___ rpm.

_____ **17.** The speed of a 20 tooth gear driven by a 6 tooth drive gear that operates at 60 rpm is ___ rpm.

_____ **18.** The number of teeth required on a driven gear required to produce 50 rpm from a drive gear having 16 teeth at 100 rpm is ___.

Vibration 16
TEST 1

Name _____ Date _____

Industrial Mechanics

_____ 1. ___ is a continuous periodic change in displacement with respect to a fixed reference.

_____ 2. Resonance is the magnification of vibration and its noise by ___% or more.
   A. 0
   B. 10
   C. 20
   D. none of the above

T F 3. All objects on earth are constantly experiencing vibration.

T F 4. Machines vibrate even when in the best operating condition.

T F 5. A vibration cycle is the complete movement from beginning to end of a vibration.

_____ 6. ___ is the location (within tolerance) of an axis of a coupled machine shaft relative to another.

_____ 7. ___ is the absolute value from a zero point (neutral) to the maximum travel on a waveform.

_____ 8. A(n) ___ is a graphic presentation of an amplitude as a function of time.

_____ 9. ___ is a measurement of frequency equal to 1 cps.

_____ 10. ___ is the position of a vibrating part at a given moment with reference to another vibrating part at a fixed reference point.

_____ 11. The peak value of acceleration is measured in units of g peak, where 1 g is equal to ___ ips^2.

_____ 12. A(n) ___ is a device that converts a physical quantity into another quantity.

_____ 13. A(n) ___ current is an electric current that is generated and dissipated in a conductive material in the presence of an electromagnetic field.

T F 14. Vibration may occur only from North to South.

T   F   **15.** A logarithmic scale is an amplitude or frequency displayed in powers of 100.

_____   **16.** Time ___ is the amplitude as a function of time.

_____   **17.** ___ is the measurement of the distance (amplitude) an object is vibrating.

_____   **18.** A(n) ___ is a multiple of a running speed (rpm) frequency.

_____   **19.** Transducers used to measure radial vibration must be attached within ___″ of the bearing.

_____   **20.** More than ___% of all rotary equipment failures are related to vibration.

_____   **21.** ___ is the square root of the sum of a set of squared instantaneous values.

_____   **22.** Frequency ___ is the amplitude versus frequency spectrum observed on an FFT analyzer.

T   F   **23.** Vibration significantly reduces the expected life of bearings and rotating shaft seals.

T   F   **24.** A change in the vibration signature of a machine indicates the ending of a defect.

T   F   **25.** Linear amplitude spectra are amplitude signals displayed in powers of 10.

_____   **26.** A(n) ___ is a device that limits vibration signals so only a single frequency or group of frequencies can pass.

_____   **27.** ___ is a graphic display used for interpretation of machine characteristics.

_____   **28.** Oil ___ is the buildup and resistance of a lubricant in a rolling-contact bearing that is rotating at excessive speeds.

T   F   **29.** Piezoelectric is the production of electricity by applying pressure to a crystal.

T   F   **30.** The magnitude of vibrations felt by humans is extremely small.

_____   **31.** ___ is the condition where the axes of two machine shafts are not aligned within tolerances.
　　　A. Alignment
　　　B. Displacement
　　　C. Misalignment
　　　D. Vibration movement

_____   **32.** ___ is the unbalance of weighted forces on opposing ends and sides of a rotor or armature.
　　　A. Coupling unbalance
　　　B. Equal rotor unbalance
　　　C. Opposing forces rotor unbalance
　　　D. Phase unbalance

33. ___ is the unbalance of weighted force across one side of the rotor or armature.
    A. Coupling unbalance
    B. Equal rotor unbalance
    C. Opposing forces rotor unbalance
    D. Phase unbalance

34. Vibration velocity is measured in ___.
    A. ft-lb
    B. ft/sec
    C. in.-lb
    D. in./sec

35. Displacement is best suited for frequencies between 1 cpm and ___ cpm.
    A. 600
    B. 6000
    C. 60,000
    D. 600,000

36. Displacement, velocity, and ___ are all direct measures of the severity of machine vibration.
    A. acceleration
    B. misalignment
    C. speed
    D. weight

37. A(n) ___ transducer is an electromechanical device that is constructed of a coil of wire supported by light springs.
    A. accelerometer
    B. displacement
    C. oscillator
    D. velocity

38. A displacement transducer is also known as a(n) ___.
    A. oscillator
    B. proximity pickup
    C. spectrometer
    D. vibrometer

39. A vibration ___ is a set of vibration readings resulting from tolerances and movement within a new machine.
    A. analysis
    B. foundation
    C. measurement
    D. signature

_____ 40. An FFT is a ___ Transform analyzer.
   A. Fast Fourier
   B. Free Fourier
   C. Free Form
   D. Front Fourier

_____ 41. ___ signal analyzers are used to display signals that are too small to show up on a time domain spectra.
   A. Current analysis
   B. Dynamic
   C. Multiple-source
   D. Time domain

_____ 42. A DSA can simultaneously display a vibration that is as much as ___ times greater than another displayed vibration.
   A. 50
   B. 100
   C. 500
   D. 1000

_____ 43. Estimates show that about ___% of all vibration problems are related to resonance.
   A. 20
   B. 30
   C. 40
   D. 60

_____ 44. Estimates show that about ___% of all vibration problems are the result of shaft misalignment.
   A. 20
   B. 30
   C. 40
   D. 60

_____ 45. Accelerometer transducers typically operate at frequencies between 120 cpm and ___ cpm.
   A. 600
   B. 6000
   C. 60,000
   D. 600,000

Unbalanced Vibrations

_____ 1. Coupling unbalance

_____ 2. Equal rotor unbalance

_____ 3. Opposing forces rotor balance

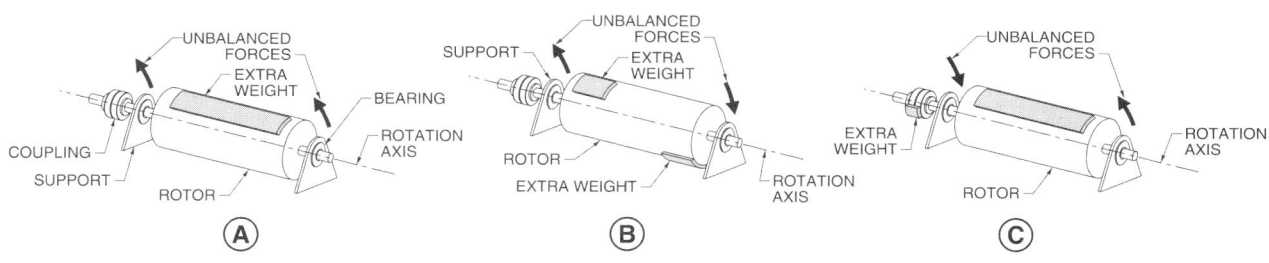

Waveform Spectrum

_____ 1. Vibration waveform

_____ 2. Peak amplitude

_____ 3. Peak-to-peak amplitude

_____ 4. 1 cycle or 1 frequency in time

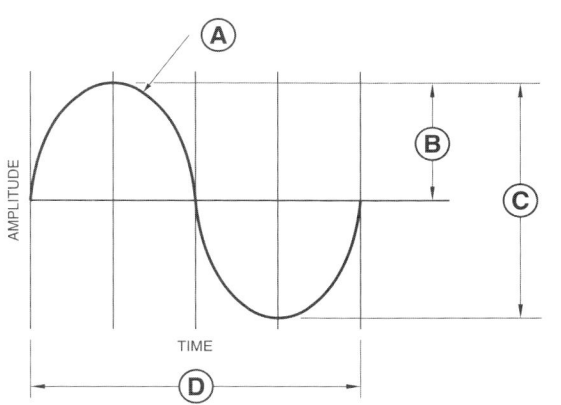

Vibration Acceleration

_____ 1. Time

_____ 2. Amplitude

_____ 3. Peak velocity

_____ 4. Peak acceleration

_____ 5. Peak-to-peak displacement

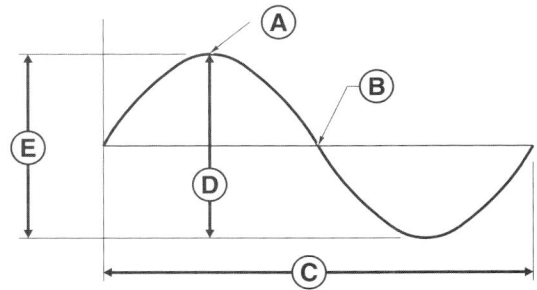

Displacement

_____ 1. Positive upper limit

_____ 2. Positive lower limit

_____ 3. Displacement

_____ 4. Peak-to-peak displacement

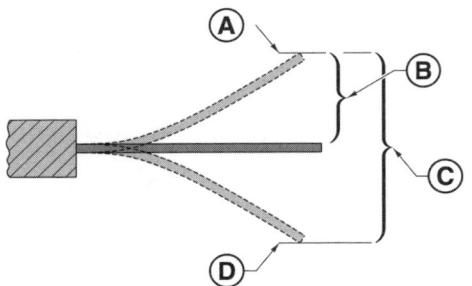

Vibration Transducers

_____ 1. Velocity

_____ 2. Accelerometer

_____ 3. Displacement

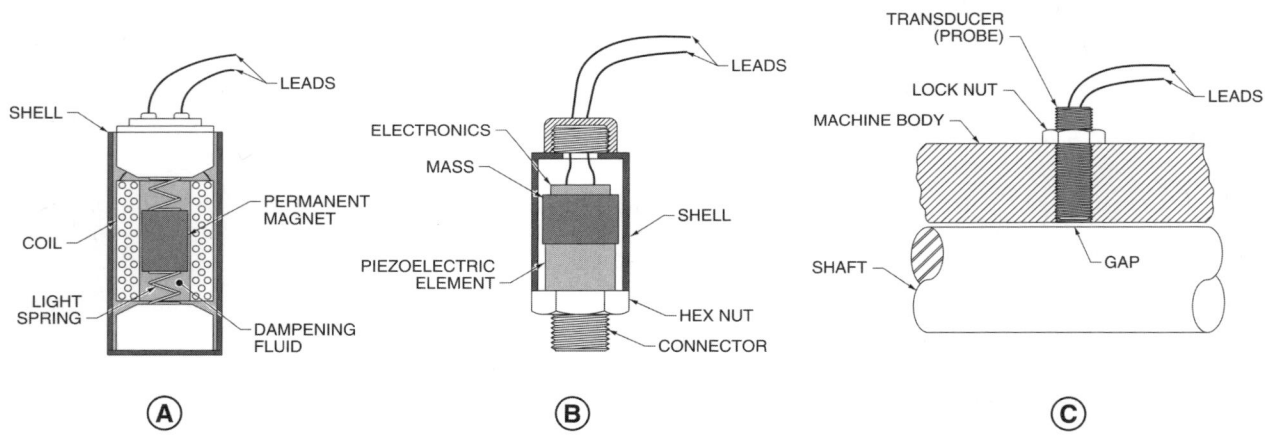

Alignment 17
TEST 1

Name _____ Date _____

Industrial Mechanics

_____ 1. ___ is the location (within tolerance) of one axis of a coupled machine shaft relative to that of another.

_____ 2. ___ is the condition where the centerlines of two machine shafts are not aligned within tolerances.

_____ 3. ___ expansion is the dimensional change in a substance due to a change in temperature.

_____ 4. A(n) ___ is a device that connects the ends of rotating shafts.

_____ 5. ___ is the process of pressing the start switch of a machine to determine if the machine starts when it is not supposed to start.
    A. Bumping
    B. Skipping
    C. Challenging
    D. none of the above

_____ 6. Shim stock is steel material manufactured in various thicknesses, ranging from ___″ to ___″.
    A. .0005; .0125
    B. .0005; .125
    C. .005; .125
    D. none of the above

_____ 7. A(n) ___ plate is a rigid steel support for firmly coupling and aligning two or more rotating devices.

_____ 8. ___ is any means of fastening a mechanism securely to a base or foundation.

T   F   9. Dowel effect is corrected by using machined washers 2 to 5 times thicker than the original washer.

T   F 10. Jack screws are used for machine movement only.

T   F 11. Runout is a radial variation from a true circle.

T F **12.** Always choose the combination that uses the least amount of shims or spacers where different shim or spacer combinations can be chosen.

_____ **13.** Good shim packs are ___ cut with each size printed on the shim.

_____ **14.** The top position, when using a dial indicator, is the ___ position.

_____ **15.** A(n) ___ gauge is a flat, tapered strip of metal with graduations in thousandths of an inch or millimeters marked along its length.

_____ **16.** Axial ___ is the axial movement of a shaft due to bearing and bearing housing clearance.

_____ **17.** When using the combination rim-and-face alignment method, offset misalignment in the vertical plane is checked by measuring the rim of the coupling at the ___ and ___ positions.
    A. 12:00; 3:00
    B. 12:00; 6:00
    C. 12:00; 9:00
    D. 3:00; 9:00

_____ **18.** When using the combination rim-and-face alignment method, offset misalignment in the horizontal plane is checked by measuring the rim of the coupling at the ___ and ___ positions.
    A. 12:00; 3:00
    B. 12:00; 6:00
    C. 12:00; 9:00
    D. 3:00; 9:00

_____ **19.** ___ misalignment is a condition where two shafts are parallel but are not on the same axis.

_____ **20.** ___ misalignment is a condition where one shaft is at an angle to the other shaft.

_____ **21.** ___ foot is a condition that occurs when one or more feet of a machine do not make complete contact with its base.

_____ **22.** A(n) ___ screw is a screw inserted through a block that is attached to a machine base plate allowing for ease in machine movement.

T F **23.** The objective of proper alignment is to align the shafts, not the couplings.

T F **24.** A spacer is steel material used for filling spaces ¼″ or less.

T F **25.** Precut stainless steel shims are recommended for alignment purposes.

_____ **26.** A base plate that is drilled and tapped to anchor a machine must be a minimum thickness of ___ times the root diameter of the anchoring bolts.
    A. 1½
    B. 2
    C. 2½
    D. 3

27. A(n) ___ is a device that measures the deviation from a true circular path.
    A. dial indicator
    B. eccentric
    C. jack screw
    D. shim

28. ___ soft foot exists when one machine foot is bent and not on the same plane as the other feet.
    A. Angular
    B. Induced
    C. Parallel
    D. Springing

29. ___ soft foot exists when one or two machine feet are higher than the others and parallel to the base plate.
    A. Angular
    B. Induced
    C. Parallel
    D. Springing

30. The shim stock thickness required to correct the angular misalignment in the vertical plane of a pump and motor assembly having a 3″ D coupling, a vertical angular gap of 0.033″, and an MTBS mounting hole distance of 4″ is ___″.
    A. 0.009
    B. 0.023
    C. 0.044
    D. 0.099

31. The adjustment required to correct the angular misalignment in the horizontal plane of a pump and motor assembly having a 3″ D coupling, a horizontal angular gap of 0.087″, and an MTBS mounting hole distance of 4″ is ___″.
    A. 0.0261
    B. 0.117
    C. 0.261
    D. 0.621

32. More than ___% of vibration problems are caused by misaligned machinery.
    A. 35
    B. 50
    C. 75
    D. 80

33. When aligning machinery, MTBS is an abbreviation for ___.
    A. machine to be set
    B. machine to be shimmed
    C. machine to be stabilized
    D. machine to be stopped

124 INDUSTRIAL MECHANICS WORKBOOK

_____ **34.** When aligning machinery, SM is an abbreviation for ___.
  A. set machine
  B. stabilized machine
  C. stationary machine
  D. stopped machine

_____ **35.** Feeler gauges can determine the air gap between two solids to accuracy within ___ of an inch.
  A. tenths
  B. hundredths
  C. thousandths
  D. ten-thousandths

Soft Foot

_____ 1. Angular

_____ 2. Parallel

_____ 3. Springing

_____ 4. Induced

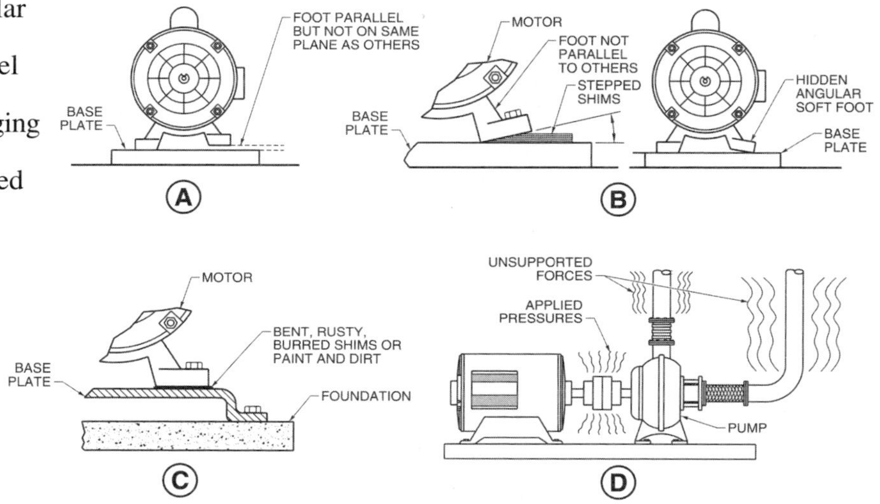

Misalignment

_____ 1. Offset

_____ 2. Angular

_____ 3. Offset and angular

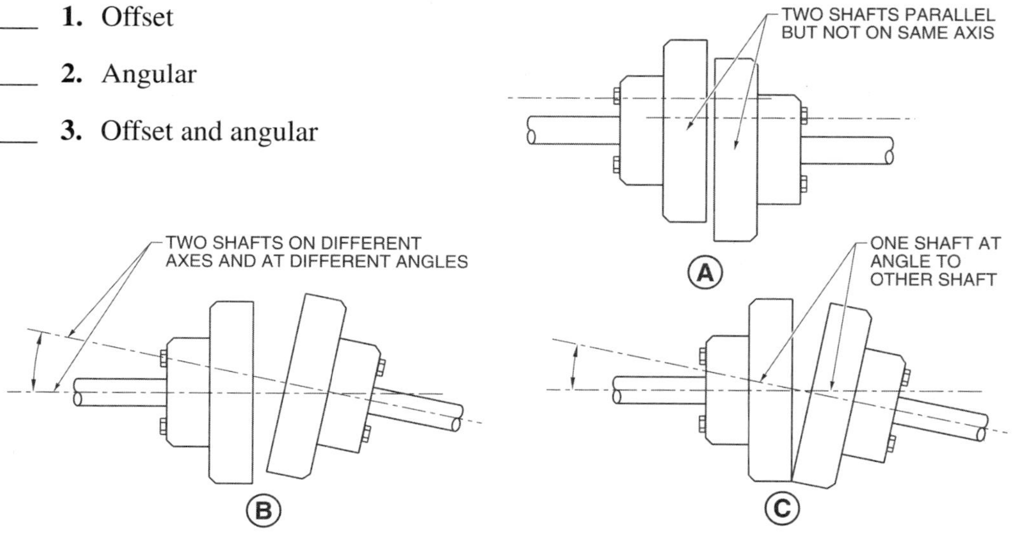

Shaft Runout

_____ 1. Bent shaft

_____ 2. Eccentric circular path

_____ 3. Poorly machined shaft

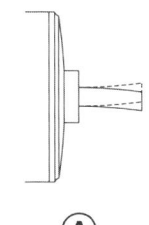 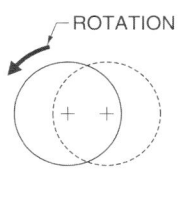 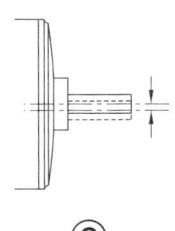

Dial Indicator Readings

_____ 1. The dial indicators at A show a TIR of ___".

_____ 2. The dial indicators at B show a TIR of ___".

_____ 3. The dial indicators at C show a TIR of ___".

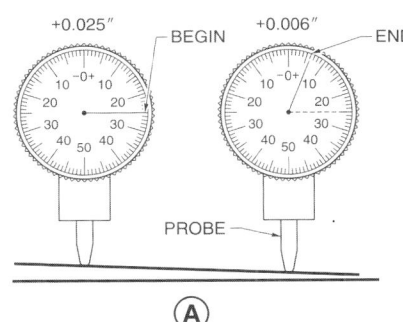 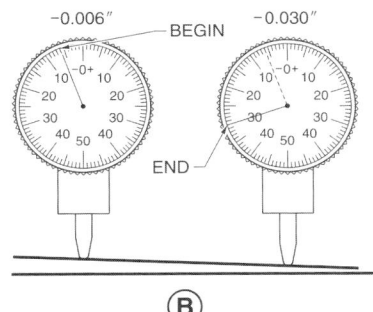 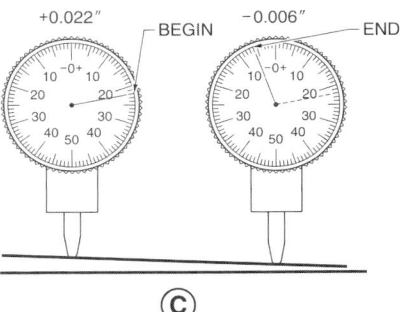

Anchoring Characteristics

_____ 1. Proper anchoring
_____ 2. Bolt bound
_____ 3. Excess bolt body
_____ 4. Bolt bottoms out
_____ 5. Dowel effect

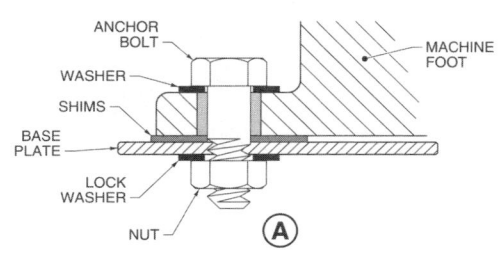
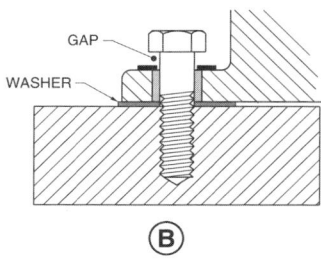
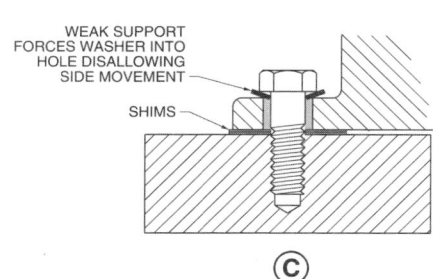
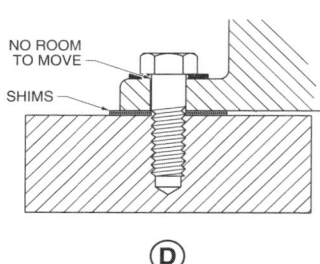
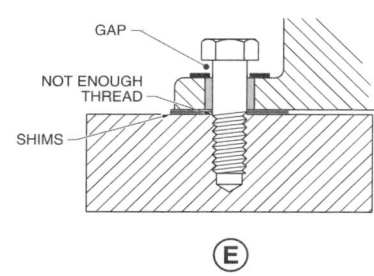

Alignment Methods

_____ 1. Straightedge
_____ 2. Rim-and-face
_____ 3. Reverse Dial
_____ 4. Electronic reverse dial
_____ 5. Laser rim-and-face

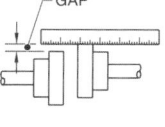
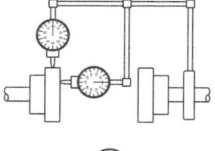
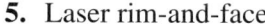
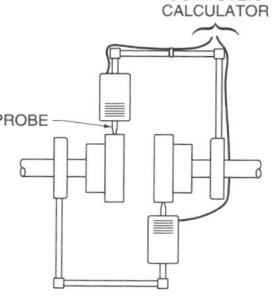
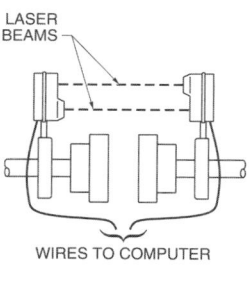

Electricity 18

TEST 1

Name _____ Date _____

Industrial Mechanics

_____ 1. ___ electricity is the accumulation of charge.

_____ 2. ___ electricity is electron flow from one atom to another.

_____ 3. A(n) ___ shell is the outermost shell of an atom.

_____ 4. ___ is the amount of electrical pressure in a circuit.

_____ 5. ___ law is the relationship between voltage, current, and resistance in a circuit.

_____ 6. A(n) ___ is a device that attracts iron and steel because of the molecular alignment of its material.

_____ 7. ___ current is a flow of electrons in only one direction.

_____ 8. ___ is the connection of all exposed noncurrent-carrying metal parts to the earth.

_____ 9. A(n) ___ tester is a device that indicates if a circuit is open or closed.

_____ 10. A(n) ___ is a test tool used to measure two or more electrical values.

_____ 11. A(n) ___ is an electric device that uses electromagnetism to change AC voltage from one level to another.

T   F   12. Fuses or circuit breakers may be bimetallic.

T   F   13. A fault current as low as 4 mA to 6 mA will activate a GFCI and interrupt the circuit.

T   F   14. Lightning is the number one cause of fires to property.

T   F   15. Resistance is the opposition to electron flow.

_____ 16. Magnetic ___ lines are the invisible lines of force that make up a magnetic field.

_____ 17. A(n) ___ is a device that converts mechanical energy into electrical energy.

_____ 18. ___ current is a flow of electrons that reverses its direction of flow at regular intervals.

_____ 19. Power ___ is the process of delivering electrical power to where it is needed.

_____ 20. The ___ is responsible for enforcing the NEC®.

_____ 21. A(n) ___ is an overcurrent device with a fusible link that melts and opens the circuit of an overcurrent condition.

T   F   22. A circuit breaker is a device with a mechanical mechanism that may manually or automatically open a circuit when an overload condition or short circuit occurs.

T   F   23. A switch is open when it allows current to flow in a circuit.

T   F   24. A solenoid is a device that converts electrical energy to a linear, mechanical force.

_____ 25. A(n) ___ is an electrical device that protects personnel by detecting potentially hazardous ground faults and quickly discontinuing power from the circuit.

_____ 26. Common insulators include rubber, plastic, air, glass, and ___.
　　A. flexible metal
　　B. gel materials
　　C. paper
　　D. water

_____ 27. An area in a food processing plant that is used for handling pulverized sugar and cocoa is an example of a Class ___, Division ___ hazardous location.
　　A. I, 1
　　B. I, 2
　　C. II, 1
　　D. II, 2

_____ 28. An area in a textile mill used for processing rayon or cotton is an example of a Class ___, Division ___ hazardous location.
　　A. I, 1
　　B. II, 1
　　C. II, 2
　　D. III, 1

_____ 29. An area in a paint manufacturing facility that contains open tanks of paint solvents such as toluene or mineral spirits is an example of a Class ___, Division ___ hazardous location.
　　A. I, 1
　　B. II, 1
　　C. I, 2
　　D. III, 1

_____ 30. The range for most voltage testers is between ___ V and 600 V.
　　A. 50
　　B. 90
　　C. 115
　　D. 208

_____ 31. A(n) ___ conductor is a wire that carries current from one side of the load to ground.

_____ 32. The ___ winding is the output or load winding of a transformer and is connected to the load.

_____ 33. A(n) ___ is an auxiliary contact used to maintain current flow to the coil of a relay.

_____ 34. The current in a 115 VAC circuit that has resistance of 48 Ω is ___ A.

_____ 35. The resistance in a 120 VAC circuit that has current of 3.47 A is ___ Ω.

Hydrogen Atom

_____ 1. Shell

_____ 2. Electron

_____ 3. Proton

_____ 4. Neutron

_____ 5. Nucleus

Hazardous Locations

_____ 1. Class I

_____ 2. Class II

_____ 3. Class III

_____ 4. Division I

_____ 5. Division II

A. Hazardous location in which hazardous substance is not normally present in air in sufficient quantities to cause an explosion or ignite hazardous materials.

B. Sufficient quantities of flammable gases and vapors present in air to cause an explosion or ignite hazardous materials.

C. Sufficient quantities of combustible dust are present in air to cause an explosion or ignite hazardous materials.

D. Easily-ignitable fibers or flyings are present in air, but not in a sufficient quantity to cause an explosion or ignite hazardous materials.

E. Hazardous location in which hazardous substance is normally present in air in sufficient quantities to cause an explosion or ignite hazardous materials

Testing Contactors and Motor Starters

_____ 1. Connect the voltage tester at A to check the incoming voltage.

_____ 2. Connect the voltage tester at B to check the control voltage.

_____ 3. Connect the voltage tester at C to check the output voltage.

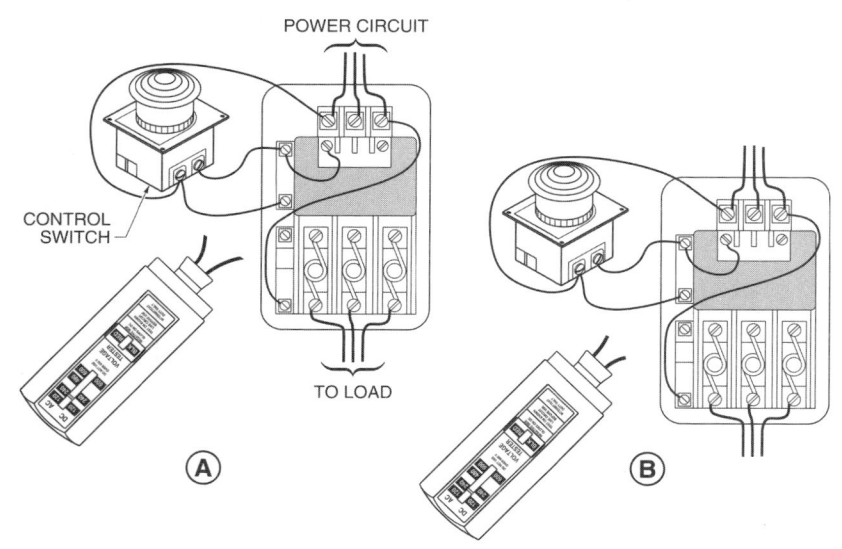
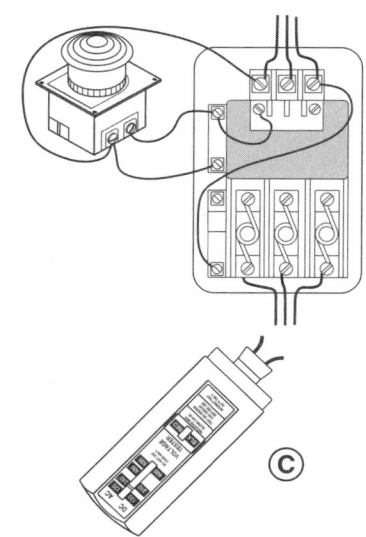

Testing Solenoids

_____ 1. Reading if coil is normal.

_____ 2. Reading if coil has a broken wire.

_____ 3. Reading if coil is shorted.

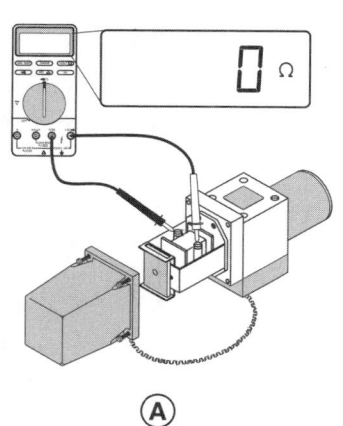
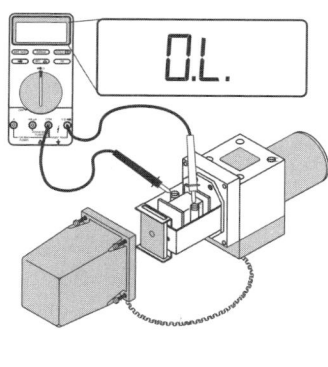
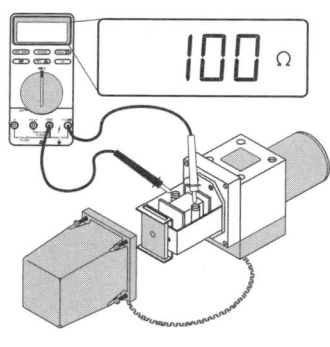

Effect of Electric Current

_____ 1. Current in 100 W lamp can electrocute 20 adults

_____ 2. Heart convulsions; usually fatal

_____ 3. Painful shock; inability to let go

_____ 4. Safe values

_____ 5. No sensation

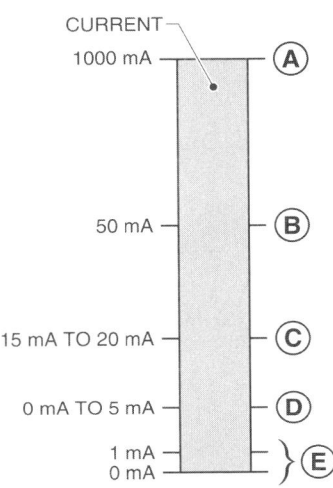

Analog Multimeters

_____ 1. Measure AC voltage

_____ 2. Measure +DC voltage and AC and DC

_____ 3. Measure –DC voltage and resistance

_____ 4. Ohm scale

_____ 5. Pointer

_____ 6. Range switch

_____ 7. Zero adjust

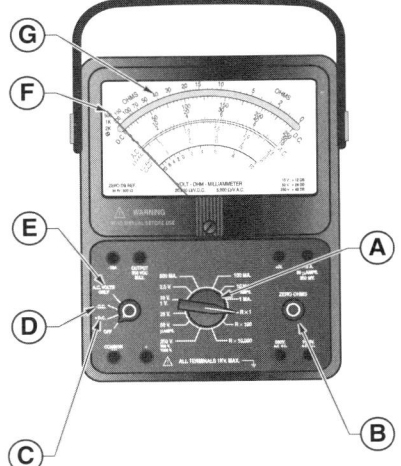

132 INDUSTRIAL MECHANICS WORKBOOK

Digital Multimeters

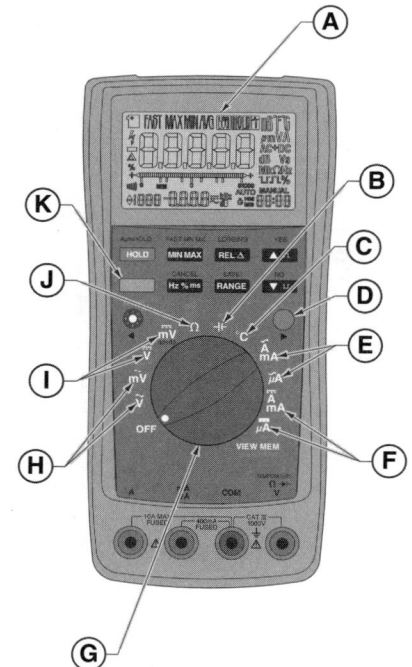

_____ 1. Access blue functions

_____ 2. Access yellow functions

_____ 3. Digital display

_____ 4. Function switch

_____ 5. Measure AC

_____ 6. Measure AC and DC

_____ 7. Measure AC voltage

_____ 8. Measure AC and DC voltage

_____ 9. Measure capacitance

_____ 10. Measure resistance

_____ 11. Measure temperature

Electrical Safety

Write five electrical safety rules that should be practiced by all personnel working with electricity.

1.

2.

3.

4.

5.

FINAL EXAM

Name _____ Date _____

Industrial Mechanics

  T F   **1.** All circles contain 360°.

  T F   **2.** Sealed bearings should be relubricated on a regularly-scheduled basis.

  T F   **3.** Torque is the twisting (rotational) force of a shaft.

  T F   **4.** The velocity of a fluid decreases as the cross-sectional area of a pipe increases.

  T F   **5.** Polygons are named according to their number of sides.

_____ **6.** A(n) ___ is a fluid that can flow readily and assume the shape of its container.

_____ **7.** A(n) ___ is a machine part that supports another part, such as a shaft, which rotates or slides in or on it.

_____ **8.** Area is always expressed in ___ units.
    A. square
    B. cubic
    C. either A or B
    D. none of the above

  T F   **9.** Angles are measured in degrees, minutes, and seconds.

_____ **10.** Lubricants are used to ___.
    A. reduce friction
    B. prevent wear
    C. prevent corrosion
    D. all of the above

_____ **11.** ___ metals are metals containing iron.

  T F   **12.** Backlash is the play between mating gear teeth.

_____ **13.** The top position, when using a dial indicator, is the ___ position.

_____ **14.** ___ is hoisting equipment or machinery by mechanical means.

_____ 15. ___ is a measure of a component's or system's useful output energy.
   A. Rate
   B. Percentage
   C. Efficiency
   D. Value

T   F   16. Machines vibrate even when in the best operating condition.

_____ 17. ___ is the three-dimensional size of an object measured in cubic units.

_____ 18. ___ is the twisting force of a shaft.
   A. Rotation
   B. Torque
   C. Shearing
   D. Bending

_____ 19. A(n) ___ load is a load in which one-half of the load is a mirror image of the other half.

_____ 20. Fixed ladders are installed in a preferred pitch range between ___° and 90° from horizontal.
   A. 45
   B. 60
   C. 75
   D. none of the above

_____ 21. ___ is the location (within tolerance) of one axis of a coupled machine shaft relative to that of another.

_____ 22. ___ energy is the energy of motion.

_____ 23. ___ current is a flow of electrons that reverses its direction of flow at regular intervals.

_____ 24. ___ is the process of maintaining a fluid film between solid surfaces to prevent their physical contact.

_____ 25. Metal ladders should not be used within ___′ of electrical circuits or equipment.

_____ 26. A binary system has ___ value(s).
   A. no
   B. one
   C. two
   D. any number of

_____ 27. ___ is the branch of science that deals with the practical application of water or other liquids at rest or in motion.

T   F   28. Walking requires friction between the feet and floor in order to move.

T   F   29. A square foot contains 12 sq in.

_____ 30. The diameter of wire rope is determined by the largest possible ___ dimension.

_____ 31. One horsepower is the amount of energy required to lift ___ lb 1′ in 1 min.
    A. 330
    B. 550
    C. 33,000
    D. 55,000

_____ 32. A(n) ___ is the smallest building block of matter than cannot be divided into smaller units without changing its basic character.

_____ 33. ___ V-belts are designated as A, B, C, D, or E.

_____ 34. ___ is a rope's attempt to rotate and untwist its strand lays while under stress.

_____ 35. A(n) ___ is a mathematical equation that contains a fact, rule, or principle.

_____ 36. The weight of the atmosphere at sea level is ___ psia.

T   F   37. The total amount of moisture that air is capable of holding varies based on the temperature of the air.

T   F   38. The objective of proper alignment is to align the shafts, not the couplings.

T   F   39. For optimum efficiency, a V-belt should touch the bottom of the pulley.

T   F   40. A 10 weight oil is thicker than a 40 weight oil.

_____ 41. A(n) ___ is a closed path through which hydraulic fluid flows or may flow.

_____ 42. ___ lift is the height to which atmospheric pressure causes a column of fluid to rise above the supply to restore equilibrium.

T   F   43. Bearings should never be struck with a hammer.

T   F   44. There are 60′ in one degree.

_____ 45. Guardrails on scaffolds must be installed no less than ___″ or more than ___″ high, with a midrail.
    A. 24; 30
    B. 30; 36
    C. 36; 42
    D. 42; 48

T   F   46. Electric motors are less efficient than air motors.

_____ 47. Atoms combine to form ___.
    A. protons
    B. particles
    C. molecules
    D. none of the above

_____ 48. ___ is a measurement of frequency equal to 1 cps.

_____ 49. A mercury ___ is an instrument that measures atmospheric pressure using a column of mercury.

T   F   50. A straight angle always contains 90°.

T   F   51. Ambient temperature is the temperature of the air surrounding a piece of equipment.

_____ 52. A pressure gauge reads ___ psig at normal atmospheric pressure.

_____ 53. ___ pistons move forward and backward alternately.

T   F   54. A switch is open when it allows current to flow in a circuit.

_____ 55. A Bourdon tube is a hollow metal tube made of brass or similar material and is ___.
   A. elliptical in cross-sectional area
   B. bent in a C-shape
   C. both A and B
   D. none of the above

_____ 56. During startup of a machine, oil ___.
   A. is cool
   B. does not flow easily
   C. A and B
   D. none of the above

_____ 57. Absolute ___ is the temperature at which substances possess no heat.

T   F   58. A machine should never be grounded by connecting a wire from the machine to a gas or oil pipe.

_____ 59. ___ misalignment is a condition where two shafts are parallel but are not on the same axis.

_____ 60. The ___ is the side of a right triangle opposite the right angle.

T   F   61. The higher the mesh number of a strainer, the smaller the opening.

_____ 62. ___ is the measurement of the distance (amplitude) an object is vibrating.

T   F   63. A person should always face the ladder when ascending or descending.

T   F   64. A better finish on a bearing produces less friction.

_____ 65. The ___ is the balancing point of a load.

_____ 66. ___ is the branch of science that deals with the transmission of energy using a gas.

T   F   67. In a bevel gear, the drive gear is the smaller gear.

_____ 68. A(n) ___ is a toothed machine element used to transmit motion between rotating shafts.

T   F   69. Static energy is the energy of motion.

_____ 70. The ___ is responsible for enforcing the NEC®.

T   F   71. Dies have a side with a 90° chamfer to prevent die-tooth breakage and to allow for a gentle cutting start.

T   F   72. Some types of power drills, such as hammer drills, can drill at speeds up to 3000 rpm and simultaneously hammer into the material at up to 50,000 blows per minute.

_____ 73. ___ drawings are often used in installation and operational manuals to show where to connect external wires and position indicating lamps, switches, and displays.
    A. Application
    B. Orthographic
    C. Location
    D. none of the above

_____ 74. A ___ is a permissible deviation from a given value or dimension.
    A. variation
    B. tolerance
    C. rule
    D. none of the above

T   F   75. Angles are measured in inches or centimeters.

_____ 76. Depth micrometers are used to measure dimensions of workpieces that have critical inside dimensions, such as ___.
    A. splines
    B. filter housings
    C. sleeve bearings
    D. all of the above

_____ 77. ___ is the verification of graduations and incremental values of a precision measuring instrument for accuracy and adjustments.
    A. Tolerance
    B. Caliper adjustment
    C. Calibration
    D. Maintenance

_____ 78. A(n) ___ is a measuring tool divided into even increments.

T   F   79. A machinist's steel protractor is a tool used to measure or mark angle measurements on flexible, pliable workpieces.

_____ 80. File parts include the point, edge, face, heel, and ___.
   A. handle
   B. head
   C. tang
   D. blade

_____ 81. Title block information typically includes ___.
   A. subject or sheet contents
   B. project title and location
   C. print division and print number
   D. all of the above

_____ 82. A ___ is a tool for measuring and laying out angles.
   A. caliper
   B. rule
   C. protractor
   D. micrometer

_____ 83. A(n) ___ shows how the individual parts of an object work together.

_____ 84. Micrometers are used for verification of component dimensions such as ___.
   A. thickness
   B. diameter
   C. depth
   D. all of the above

_____ 85. A(n) ___ is the fixed measuring surface of a micrometer.
   A. anvil
   B. thimble
   C. stop
   D. spindle

_____ 86. A ___ is usually one tool with a combination of different measuring devices attached to a steel rule and is sometimes referred to as a combination square set.
   A. caliper
   B. rule
   C. protractor
   D. reversible protractor

_____ 87. A floor plan is a plan view looking down at a building from approximately ___′ above the floor.
   A. 5
   B. 10
   C. 50
   D. 100

88. Some ___ resemble reverse-threaded screws, while others resemble square tapered rods with chiseled edges.
    A. screw extractors
    B. vises
    C. tape rules
    D. mechanical pullers

_____ 89. A(n) ___ micrometer is a micrometer with a digital electronic indicating gauge.

_____ 90. A(n) ___ drawing is a type of drawing that is used to indicate how to do work using the simplest and/or safest method.
    A. assembly
    B. sectional
    C. instructional
    D. detail

_____ 91. ___ are used with a tap wrench to "pull" the tap into a workpiece.
    A. Taps
    B. Pliers
    C. Files
    D. Mechanical pullers

_____ 92. ___ plan designs include information about the landscaping that a specific piece of property can have.
    A. Floor
    B. Foundation
    C. Structural
    D. Utility

_____ 93. Conventional protractors are ___ in shape and have an outer scale, inner scale, zero edge, and center mark.
    A. circular
    B. semicircular
    C. rectangular
    D. oblong

_____ 94. Depth micrometer measuring rod extensions typically vary by exactly ___" and are calibrated by the set manufacturer.
    A. ¼
    B. ½
    C. 1
    D. 2

_____ 95. Micrometers with a vernier scale are capable of taking measurements to the nearest ___".
    A. 0.01
    B. 0.001
    C. 0.0001
    D. 0.00001

_____ 96. A(n) ___ is additional information that is included with a set of prints.

_____ 97. Plans are two-dimensional drawings designed to indicate the ___ of objects.

_____ 98. A ___ drawing is a three-dimensional drawing that resembles a picture.
   A. head-on
   B. location
   C. detail
   D. pictorial

_____ 99. ___ are used to locate the centers of windows, doors, and electrical enclosures, and to indicate that an object is round or cylindrical in shape.
   A. Dimension lines
   B. Title blocks
   C. Centerlines
   D. all of the above

_____ 100. Electronic calipers can indicate readings in thousandths of an inch or ___ of a millimeter.
   A. tenths
   B. fiftieths
   C. hundredths
   D. thousandths

_____ 101. A micrometer used to measure grooves or keyways has a ___-shaped anvil and spindle.
   A. ball
   B. disc
   C. point
   D. cup

_____ 102. Underground utilities are commonly indicated with a ___ line or with a solid line that is broken for placement of a letter indicting the type of utility.
   A. dashed
   B. dotted
   C. dotted and dashed
   D. solid

_____ 103. Single-cut files have a single set of teeth and are used to make cuts at an angle between ___° and ___°.
   A. 45, 90
   B. 65, 85
   C. 75, 85
   D. 75, 90

104. ___ are designed with standard micrometer heads (thimble and barrel) attached to a flat "tee" base.
   A. Inside micrometers
   B. Outside micrometers
   C. Calipers
   D. none of the above

105. A ___ is a reproduction of original drawings created by an architect or engineer.
   A. print
   B. plan
   C. legend sheet
   D. note

106. ___ taps are used after a taper tap has been used to start a true and straight thread.
   A. Bottom
   B. Plug
   C. Die
   D. Wrench

107. The main parts of a handsaw are the blade, teeth, back, and ___.
   A. chisel
   B. head
   C. handle
   D. none of the above

108. General-purpose section lines are typically drawn on a 45° angle and ___″ apart.
   A. 1/32
   B. 1/16
   C. 1/10
   D. 1/8

109. End-cutting pliers are used for cutting ___ close to the workpiece.
   A. wire
   B. nails
   C. rivets
   D. all of the above

110. A(n) ___ is a cut that is made against the direction of the wood grain and is made with full, even strokes at about a 45° angle.

Problems

_____ 1. The area of Circle A is ___ sq in.

_____ 2. The horsepower required to lift load B is ___ HP.

_____ 3. The area of Surface A on Block C is ___ sq in.

_____ 4. The volume of Block C is ___ cu in.

_____ 5. ___ lb of effort force is required to lift the resistance force of Fulcrum D.

_____ 6. The horsepower required to lift the 6 t load is ___ HP.

_____ 7. The temperature on the Fahrenheit scale equals ___°R.

_____ 8. The driven pulley speed at E is ___ rpm.

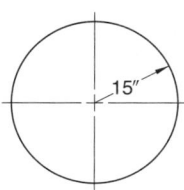

CIRCLE A

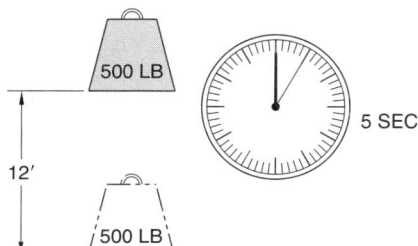

LOAD B

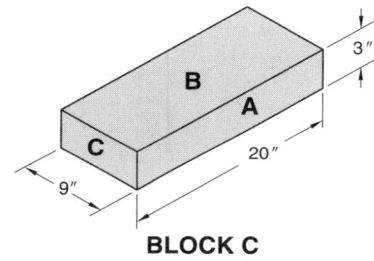

BLOCK C

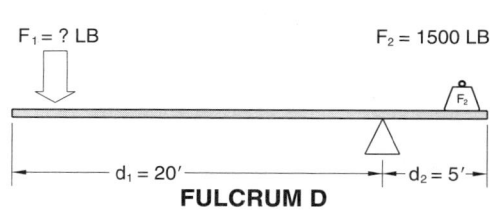

FULCRUM D

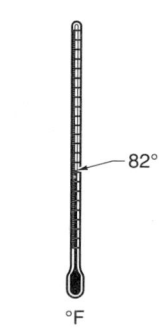

°F

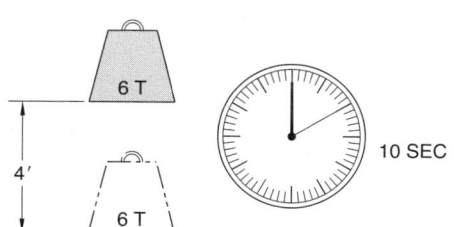

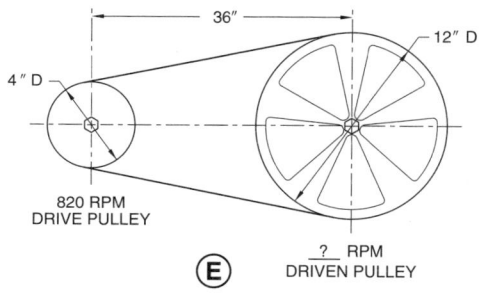

APPENDIX

Weight of Steel and Brass Bar Stock 143
Weight of Steel Plate 143
Lead Line Factors 144
Sling Eyebolt Capacity Loss 144
Sling Rope Load Capacity 6 × 19 Classification (2000 lb Ton) 144
Vertical Sling Component Load Capacity 6 × 19 IPS-FC Classification (2000 lb Ton) 144
Sling Angle Loss Factors 144
Rope Bending Efficiency 144
Wire Rope Strength 145
Pole Scaffold Components 145
Sling Vertical Capacities 146
Round Sling Color and Capacity Rating 146
Atmospheric Pressure vs Lift 147
Extension Ladder Section Overlap 147
Choker Hitch Capacities 147
Fluid Weights/Temperature Standards 147
Sling Material Strength Capacities 147
Angle Positioning 147
Formulas 148-150
Motor Horsepower 151
Three-Phase Voltage Values 152
Ohm's Law and Power Formula 152
Power Formula Abbreviations and Symbols 152
Power Formulas – 1ϕ, 3ϕ 152
Horsepower to Torque Conversion 153
Hoisting Equipment Checklist 154

WEIGHT OF STEEL AND BRASS BAR STOCK*

Diameter or Thickness**	Round Steel	Square Steel	Brass
1/4	.167	—	.181
1/2	.667	—	.724
3/4	1.50	—	1.63
1	2.67	3.4	2.89
1 1/4	4.17	—	4.52
1 1/2	6.01	7.7	6.51
1 3/4	8.18	—	8.86
2	10.68	—	11.57
4	42.7	54.4	—
5	66.8	85.0	—
6	96.1	122.4	—
10	267.0	340.0	—
12	384.5	489.6	—

* in lb/ft
** in in.

WEIGHT OF STEEL PLATE*

Thickness**	Weight
1/16	2.55
1/8	5.1
3/16	7.65
1/4	10.2
5/16	12.75
3/8	15.3
1/2	20.4
5/8	25.5
3/4	30.6
1	40.8
1 1/4	51.0
1 1/2	61.2
2	81.6

* in lb/sq ft
** in in.

LEAD LINE FACTORS*

Parts of Line	Plain Bearing Pulleys	Rolling-Contact Bearing Pulleys
1	1.09	1.04
2	.568	.530
3	.395	.360
4	.309	.275
5	.257	.225
6	.223	.191
7	.199	.167
8	.181	.148
9	.167	.135
10	.156	.123
11	.147	.114
12	.140	.106
13	.133	.100
14	.128	.095
15	.124	.090

* based on equal number of pulleys

SLING EYEBOLT CAPACITY LOSS

Sling Angle*	Capacity Reduction**
90	100
60 – 89	70
45 – 59	30
Less than 45	25

* in degrees
** in %

SLING ROPE LOAD CAPACITY 6 x 19 CLASSIFICATION (2000 LB TON)

Rope Dia*	Choker	Vertical Load	2-Leg 30°	2-Leg 45°	2-Leg 60°
$\frac{1}{4}$	.35	.65	.58	.50	.31
$\frac{3}{8}$	.84	1.8	1.68	1.37	.95
$\frac{1}{2}$	1.50	2.50	2.96	2.41	1.71
$\frac{3}{4}$	3.20	6.0	6.58	5.37	3.80
1	5.5	10.0	11.56	9.44	6.58

* in in.

VERTICAL SLING COMPONENT LOAD CAPACITY 6 x 19 IPS-FC CLASSIFICATION* (2000 LB TON)

Rope Dia**	Spelter/Swaged	U-Bolt	Wedge	Mechanical Splice
$\frac{1}{4}$	.54	.43	.43	.49
$\frac{3}{8}$	1.22	.97	.97	1.09
$\frac{1}{2}$	2.14	1.71	1.71	1.92
$\frac{3}{4}$	4.76	3.80	3.80	4.28
1	8.36	6.68	6.68	7.52

* rates include safety factor of 5
** in in.

SLING ANGLE LOSS FACTORS

Angle from Horizontal*	Loss Factor
90	1.000
85	.996
80	.985
75	.966
70	.940
65	.906
60	.866
55	.819
50	.766
45	.707
40	.643
35	.574
30	.500

* in degrees

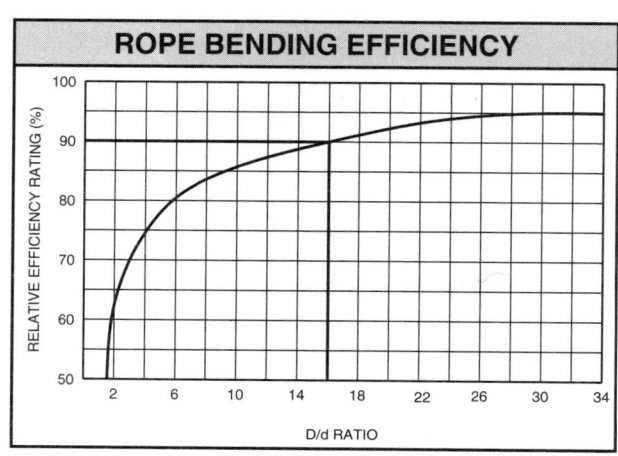

WIRE ROPE STRENGTH

Nominal Diameter*	Classification	Nominal Breaking Strength per 2000 lb Ton	
		IPS**	EIPS†
$\frac{1}{4}$	6 x 19 STANDARD HOISTING FIBER CORE	2.74	—
	6 x 19 STANDARD HOISTING IWRC	—	3.40
	8 x 19 SPECIAL FLEXIBLE HOISTING FIBER CORE	2.35	—
	18 x 7 NONROTATING	2.51	—
$\frac{3}{8}$	6 x 19 STANDARD HOISTING FIBER CORE	6.10	—
	6 x 19 STANDARD HOISTING IWRC	—	7.55
	8 x 19 SPECIAL FLEXIBLE HOISTING FIBER CORE	5.24	—
	18 x 7 NONROTATING	5.59	—
$\frac{1}{2}$	6 x 19 STANDARD HOISTING FIBER CORE	10.7	—
	6 x 19 STANDARD HOISTING IWRC	—	13.3
	8 x 19 SPECIAL FLEXIBLE HOISTING FIBER CORE	9.23	—
	18 x 7 NONROTATING	9.85	—

STANDARD HOISTING 6 x 19 SEALE WITH FIBER CORE

SPECIAL FLEXIBLE HOISTING 8 x 19 WARRINGTON WITH FIBER CORE

NONROTATING WIRE ROPE 18 x 7 WITH FIBER CORE

* in in.
** IPS - improved plow steel
† EIPS - extra improved plow steel

POLE SCAFFOLD COMPONENTS*

Type	Poles	Bearers	Ledgers (Stringers)	Braces	Planking	Rails
Light-duty** single-pole	20′ or less – 2 × 4 60′ or less – 4 × 4	3′ width – 2 × 4 5′ width – 4 × 4	20′ or less – 1 × 4 60′ or less – 1¼ × 9	1 × 4	2 × 10	2 × 4
Medium-duty† single-pole	60′ or less – 4 × 4	2 × 10	2 × 10	1 × 6	2 × 10	2 × 4
Heavy-duty‡ single-pole	60′ or less – 4 × 4	2 × 10	2 × 10	2 × 4	2 × 10	2 × 4
Light-duty* double-pole	20′ or less – 2 × 4 60′ or less – 4 × 4	3′ width – 2 × 4 5′ width – 4 × 4	20′ or less – 1¼ × 4 60′ or less – 1¼ × 9	1 × 4	2 × 10	2 × 4
Medium-duty† double-pole	60′ or less – 4 × 4	2 × 10	2 × 10	1 × 6	2 × 10	2 × 4
Heavy-duty‡ double-pole	60′ or less – 4 × 4	2 × 10	2 × 10	2 × 4	2 × 10	2 × 4

* all members except planking are used on edge
** not to exceed 25 lb/sq ft
† not to exceed 50 lb/sq ft

SLING VERTICAL CAPACITIES						
Width*	Class 5**			Class 7†		
	Types I, II, III, IV	Type V	Type VI	Types I, II, III, IV	Type V	Type VI
1	1100	2200	—	1600	3200	—
1½	1600	3200	—	2300	4600	—
1¾	1900	3800	—	2700	5400	—
2	2200	4400	3600	3100	6200	5800
3	3300	6600	—	4700	9400	—
3½	—	—	5800	—	—	8400
4	4400	8800	6800	6200	12,400	11,000
5	5500	11,000	—	7800	15,600	—
6	6600	13,200	10,000	9300	18,600	16,000

* in in.
** minimum certified tensile strength of 6800 lb per in. of width
† minimum certified tensile strength of 9800 lb per in. of width

ROUND SLING COLOR AND CAPACITY RATING*					
Round Sling Size No.	Color	Vertical	Choker	Vertical Basket	45° Basket
		Weight	Weight	Weight	Weight
1	Purple	2600	2100	5200	3700
2	Green	5300	4200	10,600	7500
3	Yellow	8400	6700	16,800	11,900
4	Tan	10,600	8500	21,200	15,000
5	Red	13,200	10,600	26,400	18,700
6	White	16,800	13,400	33,600	23,800
7	Blue	21,200	17,000	42,400	30,000
8	Orange	25,000	20,000	50,000	35,400
9	Orange	31,000	24,800	62,000	43,800
10	Orange	40,000	32,000	80,000	56,600
11	Orange	53,000	42,400	106,000	74,900
12	Orange	66,000	52,800	132,000	93,000

* in lb

Appendix 147

ATMOSPHERIC PRESSURE VS LIFT

Altitude above Sea Level*	Barometer Reading**	Atmospheric Pressure†	Theoretical Lift at Standard Temperature of 62°F*
0	29.92	14.7	34
1000	28.8	14.2	33
2000	27.7	13.6	31.5
3000	26.7	13.1	30.2
4000	25.7	12.6	29.1
5000	24.7	12.1	28
6000	23.8	11.7	27
7000	22.9	11.2	26
8000	22.1	10.8	25
9000	21.2	10.4	24
10,000	20.4	10.0	23

* in ft
** in in. Hg
† in psi

SLING MATERIAL STRENGTH CAPACITIES*

6 x 19

ROPE DIA**	Rated Capacities (in Tons)†		
	VERTICAL	CHOKER	BASKET
$\frac{1}{4}$	.51	.38	1.0
$\frac{5}{16}$	.79	.60	1.6
$\frac{3}{8}$	1.1	.85	2.2
$\frac{7}{16}$	1.5	1.1	3.0
$\frac{1}{2}$	2.0	1.5	4.0
$\frac{9}{16}$	2.5	1.9	5.0
$\frac{5}{8}$	3.1	2.3	6.2
$\frac{3}{4}$	4.4	3.3	8.8
$\frac{7}{8}$	6.0	4.5	12.0
1	7.7	5.9	15.0

* improved plow steel/fc
** in in.
† rates include safety factor of 5

EXTENSION LADDER SECTION OVERLAP

Ladder Length*	Overlap*
8 to 36	3
36 to 48	4
48 to 60	5

* in ft

CHOKER HITCH CAPACITIES

Angle of Choke*	Sling Rated Load Factor
120 – 180	.75
90 – 119	.65
60 – 89	.55
30 – 59	.40

* in degrees

FLUID WEIGHTS/TEMPERATURE STANDARDS

Fluid	Weight*	Temperature**
Air	4.33×10^{-5}	20°C/68°F @ 29.92 in. Hg
Gasoline	.0237 – .0249	20°C/68°F
Kerosene	.0296	20°C/68°F
Mercury	.49116	0°C/32°F
Lubricating Oil	.0307 – .0318	15°C/59°F
Fuel Oil	.0336 – .0353	15°C/59°F
Water	.0361	4°C/39°F
Sea Water	.0370	15°C/59°F

* in lb/cu in.
** laboratory temperature conditions under which numerical values are defined

ANGLE POSITIONING

Vertical Dimension	Horizontal Dimension*
8	2
10	$2\frac{1}{2}$
12	3
16	4
20	5
24	6
28	7
32	8
36	9
40	10
44	11

* in ft

FORMULAS...

AREA

Circle (Radius)
$A = 3.1416 \times r^2$
where
A = area
3.1416 = constant (π)
r^2 = radius squared

Circle (Diameter)
$A = .7854 \times D^2$
where
A = area
$.7854$ = constant
D^2 = diameter squared

Square or Rectangle
$A = l \times w$
where
A = area
l = length
w = width

Triangle
$A = \frac{1}{2}bh$
where
A = area
$\frac{1}{2}$ = constant
b = base
h = height

VOLUME

Cylinder (Radius)
$V = \pi r^2 \times l$
where
V = volume
$\pi = 3.1416$
r^2 = radius squared
l = length

Cylinder (Diameter)
$V = .7854 \times D^2 \times h$
where
V = volume
$.7854$ = constant
D^2 = diameter squared
h = height

Rectangular Solid
$V = l \times w \times h$
where
V = volume
l = length
w = width
h = height

Sphere (Radius)
$V = \frac{4\pi r^3}{3}$
where
V = volume
4 = constant
$\pi = 3.1416$
r^3 = radius cubed
3 = constant

Sphere (Diameter)
$V = \frac{\pi D^3}{6}$
where
V = volume
$\pi = 3.1416$
D^3 = diameter cubed
6 = constant

Cone
$V = \frac{A_b \, a}{3}$
where
V = volume
A_b = area of base
a = altitude
3 = constant

PYTHAGOREAN THEOREM
$c = \sqrt{a^2 + b^2}$
where
c = length of hypotenuse
a^2 = length of one side squared
b^2 = length of other side squared

TEMPERATURE

Converting Fahrenheit to Celsius
$°C = \frac{°F - 32}{1.8}$
where
$°C$ = degrees Celsius
$°F$ = degrees Fahrenheit
32 = difference between bases
1.8 = ratio between bases

Converting Celsius to Fahrenheit
$°F = (1.8 \times °C) + 32$
where
$°F$ = degrees Fahrenheit
1.8 = ratio between bases
$°C$ = degrees Celsius
32 = difference between bases

Converting Fahrenheit to Rankine
$°R = 460 + °F$
where
$°R$ = degrees Rankine
460 = difference between bases
$°F$ = degrees Fahrenheit

Converting Celsius to Kelvin
$°K = 273 + °C$
where
$°K$ = degrees Kelvin
273 = difference between bases
$°C$ = degrees Celsius

STOCK MATERIAL WEIGHT
$W = l \times w/ft$
where
W = weight (in lb)
l = length (in ft)
w/ft = weight (in lb/ft)

LIFTING CAPACITY
$LC = vl \times l \times s$
where
LC = load capacity (in t)
vl = vertical load rate (from Vertical Sling Component Load Capacity 6 × 19 IPS-FC Classification table)
l = number of sling legs (not more than two)
s = loss factor (from Sling Angle Loss Factors table)

ROPE BENDING LOAD RATING
$R_{br} = R_{lr} \times R_{eff}$
where
R_{br} = rope bending load rating
R_{lr} = rope load rating
R_{eff} = relative efficiency rating

D/d Ratio
$R = \frac{D}{d}$
where
R = D/d ratio
D = diameter of rope curve (in in.)
d = diameter of rope (in in.)

ROPE STRENGTH
$R_s = t \times 5$
where
R_s = rope strength (in t)
t = weight (in t)
5 = constant (safety factor)

TONS
$T = \frac{w}{2000}$
where
T = weight (in t)
w = weight (in lb)
2000 = constant (to convert lb to t)

...FORMULAS...

HOLDING LOADS

$$L = \frac{w}{p}$$

where
L = lead line force (in lb)
w = total load weight including weight of slings, containers, etc. (in lb)
p = number of parts

MOVING LOADS

$$L = f \times w$$

where
L = lead line force (in lb)
f = lead line factor (from Lead Line Factors table)
w = weight of load (in lb)

COMPRESSOR SIZE

$$HP = \frac{scfm}{4}$$

where
HP = horsepower
$scfm$ = standard cubic feet per minute
4 = constant

WORKING LOAD CAPACITY

$$L = \frac{c \times wl}{s}$$

where
L = working load capacity (in lb)
c = constant (.21 for sling angles less than 45°; .25 for sling angles greater than 45°)
wl = eyebolt working load limit (in lb)
s = sling angle loss factor (from Sling Angle Loss Factors table)

ABSOLUTE PRESSURE

$$psia = psig + 14.7$$

where
$psia$ = pounds per square inch absolute
$psig$ = pounds per square inch gauge
14.7 = constant (atmospheric pressure at standard conditions)

CYLINDER PRESSURE

$$P = \frac{F}{A}$$
$$F = P \times A$$
$$A = \frac{F}{P}$$

where
P = pressure
F = force
A = area

CYLINDER CAPACITY

$$C = \frac{V}{231}$$

where
C = capacity (in gal.)
V = volume (in cu in.)
231 = constant (cu in. of fluid per gal.)

PRESSURE OF FLUID IN CYLINDER

$$P = w \times h$$

where
P = pressure at base (in psi)
w = weight of fluid (in lb/cu in. from Fluid Weights/Temperature Standards table)
h = height (in in.)

FLUID VELOCITY

$$v = \frac{x_2 - x_1}{t_2 - t_1}$$

where
v = velocity (in ft/sec)
x_2 = final position (in ft)
x_1 = initial position (in ft)
t_2 = final time (in sec)
t_1 = initial time (in sec)

VELOCITY OF FLUID IN PIPE

$$v = \frac{l_2}{\frac{A \times l_1}{231} \times \frac{60}{Q}}$$

where
v = velocity (in ft/sec)
l_2 = length of pipe (in ft)
A = cross-sectional area of pipe (in sq in.)
l_1 = length of pipe (in in.)
231 = constant (cu in. of fluid per gallon)
Q = flow rate (in gpm)
60 = constant (sec in 1 min)

SPEED OF CYLINDER ROD

$$s = 231 \times \frac{Q}{.7854} \times D^2$$

where
s = speed of extension (in in./min)
231 = constant (cu in. of fluid per gallon)
Q = flow rate (in gpm)
$.7854$ = constant
D^2 = diameter of cylinder squared

FORCE TO OVERCOME RESISTANCE FORCE

$$F_1 = \frac{F_2 \times d_2}{d_1}$$

where
F_1 = effort force (in lb)
F_2 = resistance force (in lb)
d_1 = distance between effort force and fulcrum (in ft)
d_2 = distance between resistance force and fulcrum (in ft)

RESULTING FORCE WITHIN VESSEL

$$F_2 = F_1 \times \frac{A_2}{A_1}$$

where
F_2 = resulting force (in lb)
F_1 = input force (in lb)
A_2 = area of output pressure (in sq in.)
A_1 = area of input pressure (in sq in.)

EFFICIENCY

$$Eff_T = Eff_1 \times Eff_2 \times 100$$

where
Eff_T = total efficiency (in %)
Eff_1 = efficiency of component 1
Eff_2 = efficiency of component 2
100 = constant (to convert to percent)

POWER

$$P = \frac{F \times d}{t}$$

where
P = power (in lb-ft/time)
F = force (in lb)
d = distance (in ft or in.)
t = time (in sec, min, or hr)

HORSEPOWER

Mechanical

$$HP = \frac{F \times d}{550 \times t}$$

where
HP = horsepower
F = force (in lb)
d = distance (in ft)
550 = constant
t = time (in sec)

Hydraulic

$$HP = P \times Q \times .000583$$

where
HP = horsepower
P = pressure (in psi)
Q = flow rate (in gpm)
$.000583$ = constant

TORQUE

$$T = \frac{P \times d}{2\pi}$$

where
T = torque (in lb-in.)
P = pressure (in psi)
d = motor displacement (in cu in.)
π = constant (3.1416)

... FORMULAS

FINAL PRESSURE
$$P_2 = \frac{P_1 \times V_1}{V_2}$$
where
P_2 = final pressure (in psia)
P_1 = initial pressure (in psia)
V_1 = initial volume (in cubic units)
V_2 = final volume (in cubic units)

FINAL VOLUME
$$V_2 = \frac{P_1 \times V_1}{P_2}$$
where
V_2 = final volume (in cubic units)
P_1 = initial pressure (in psia)
V_1 = initial volume (in cubic units)
P_2 = final pressure (in psia)

CHARLES' LAW
$$V_2 = \frac{V_1 \times T_2}{T_1}$$
where
V_2 = final volume (in cubic units)
V_1 = initial volume (in cubic units)
T_2 = final temperature (in °R)
T_1 = initial temperature (in °R)

GAY-LUSSAC'S LAW
$$P_2 = \frac{P_1 \times T_2}{T_1}$$
where
P_2 = final pressure (in psia)
P_1 = initial pressure (in psia)
T_2 = final temperature (in °R)
T_1 = initial temperature (in °R)

COMBINED GAS LAW
$$P_2 = \frac{P_1 \times V_1}{T_1} \times \frac{T_2}{V_2}$$
where
P_2 = final pressure (in psia)
P_1 = initial pressure (in psia)
V_1 = initial volume (in cubic units)
T_1 = initial temperature (in °R)
T_2 = final temperature (in °R)
V_2 = final volume (in cubic units)

RATIO OF COMPRESSION
$$R_c = \frac{P_2}{P_1}$$
where
R_c = ratio of compression
P_2 = final pressure (in psia)
P_1 = initial pressure (in psia)

PRESSURE LOSS
$$\Delta P = \frac{CQ^2}{1000} \times \frac{14.7}{14.7 + P}$$
where
ΔP = pressure drop (in psi)
C = constant (from Pressure Loss Constants table)
Q = air flow rate (in scfm)
14.7 = constant (atmospheric pressure)
1000 = constant
P = working pressure (in psi)

BELT LENGTH
$$L = 2 \times C + 1.57 \times (D + d) + \frac{(D - d)^2}{4 \times C}$$
where
L = belt length (in in.)
2 = constant
C = distance between pulley centers (in in.)
1.57 = constant
D = large pulley diameter (in in.)
d = small pulley diameter (in in.)
4 = constant

DEFLECTION HEIGHT
$$h = L \times 1/64''$$
where
h = deflection height (in in.)
L = span length (in in.)
$1/64''$ = constant (.0156")

DRIVEN PULLEY SPEED
$$N_d = \frac{PD_m \times N_m}{PD_d}$$
where
N_d = driven pulley speed (in rpm)
PD_m = drive pulley diameter (in in.)
N_m = drive pulley speed (in rpm)
PD_d = driven pulley diameter (in in.)

TORQUE
$$T = F \times D$$
where
T = torque (in lb-ft)
F = force (in lb)
D = distance (in in. or ft)

TORQUE OF ROTATING MACHINE
$$T = \frac{5252 \times HP}{rpm}$$
where
T = torque (in lb-ft)
5252 = constant (33,000 lb-ft ÷ π × 2)
HP = horsepower
rpm = revolutions per minute

HORSEPOWER REQUIRED TO OVERCOME LOAD
$$HP = \frac{T \times rpm}{5252}$$
where
HP = horsepower
T = torque (in lb-ft)
rpm = revolutions per minute
5252 = constant (33,000 lb-ft ÷ π × 2)

SPEED OF DRIVEN GEAR
$$N_2 = \frac{T_1 \times N_1}{T_2}$$
where
N_2 = speed of driven gear (in rpm)
T_1 = number of teeth on drive gear
N_1 = speed of drive gear (in rpm)
T_2 = number of teeth on driven gear

COEFFICIENT OF FRICTION
$$f = \frac{F}{N}$$
where
f = coefficient of friction
F = force at which sliding occurs (in lb)
N = object weight (in lb)

Appendix

MOTOR HORSEPOWER

Pump Flow*	Pump Pressure†										
	100	250	500	750	1000	1250	1500	2000	3000	4000	5000
1	.07	.18	.36	.54	.72	.91	1.09	1.45	2.18	2.91	3.64
2	.14	.36	.72	1.09	1.45	1.82	2.18	2.91	4.37	5.83	7.29
3	.21	.54	1.09	1.64	2.18	2.73	3.28	4.37	6.56	8.75	10.93
4	.29	.72	1.45	2.18	2.91	3.64	4.37	5.83	8.75	11.66	14.58
5	.36	.91	1.82	2.73	3.64	4.55	5.46	7.29	10.93	14.58	18.23
8	.58	1.45	2.91	4.37	5.83	7.29	8.75	11.66	17.50	23.33	29.17
10	.72	1.82	3.64	5.46	7.29	9.11	10.93	14.58	21.87	29.17	36.46
12	.87	2.18	4.37	6.56	8.75	10.93	13.12	17.50	26.25	35.00	43.75
15	1.09	2.73	5.46	8.20	10.93	13.67	16.40	21.87	32.81	43.75	54.69
20	1.45	3.64	7.29	10.93	14.58	18.23	21.87	29.17	43.75	58.34	72.92
25	1.82	4.55	9.11	13.67	18.23	22.79	27.34	36.46	54.69	72.92	91.16
30	2.18	5.46	10.93	16.40	21.87	27.34	32.81	43.75	65.63	87.51	109.39
35	2.55	6.38	12.76	19.14	25.52	31.90	38.28	51.05	76.57	102.10	127.62
40	2.91	7.29	14.58	21.87	29.17	36.46	43.75	58.34	87.51	116.68	145.85
45	3.28	8.20	16.40	24.61	32.81	41.02	49.22	65.63	98.45	131.27	164.08
50	3.64	9.11	18.23	27.34	36.46	45.58	54.69	72.92	109.39	145.85	182.32
55	4.01	10.20	20.05	30.08	40.11	50.13	60.16	80.22	120.33	160.44	200.55
60	4.37	10.93	21.87	32.81	43.75	54.69	65.63	87.51	131.27	175.02	218.78
65	4.74	11.85	23.70	35.55	47.40	59.25	71.10	94.80	142.21	189.61	237.01
70	5.10	12.76	25.52	38.28	51.05	63.81	76.57	102.10	153.13	204.20	255.25
75	5.46	13.67	27.36	41.02	54.69	68.37	82.04	109.39	164.08	218.78	273.48
80	5.83	14.58	29.17	43.75	58.34	72.92	87.51	116.68	175.02	233.37	291.71
90	6.56	16.40	32.81	49.22	65.63	82.04	98.45	131.27	196.90	262.54	328.17
100	7.29	18.23	36.46	54.69	72.92	91.16	109.39	145.85	218.78	291.71	364.64

* in gpm
† pump pressure in psi (efficiency assumed to be 80%)

THREE-PHASE VOLTAGE VALUES
For 208 V × 1.732, use 360
For 230 V × 1.732, use 398
For 240 V × 1.732, use 416
For 440 V × 1.732, use 762
For 460 V × 1.732, use 797
For 480 V × 1.732, use 831

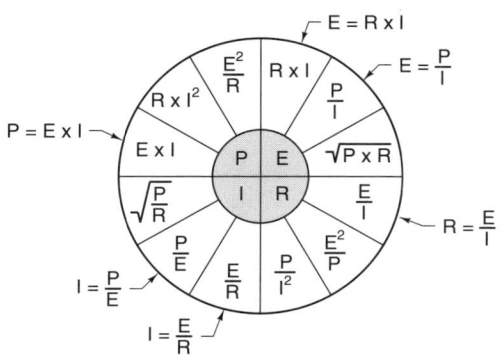

VALUES IN INNER CIRCLE ARE EQUAL TO VALUES IN CORRESPONDING OUTER CIRCLE

OHM'S LAW AND POWER FORMULA

POWER FORMULA ABBREVIATIONS AND SYMBOLS	
P = Watts	V = Volts
I = Amps	VA = Volt Amps
A = Amps	φ = Phase
R = Ohms	√ = Square Root
E = Volts	

POWER FORMULAS – 1φ, 3φ					
Phase	To Find	Use Formula	\multicolumn{3}{c}{Example}		
			Given	Find	Solution
1φ	I	$I = \dfrac{VA}{V}$	32,000 VA, 240 V	I	$I = \dfrac{VA}{V}$ $I = \dfrac{32{,}000\,VA}{240\,V}$ **I = 133 A**
1φ	VA	$VA = I \times V$	100 A, 240 V	VA	$VA = I \times V$ $VA = 100\,A \times 240\,V$ **VA = 24,000 VA**
1φ	V	$V = \dfrac{VA}{I}$	42,000 VA, 350 A	V	$V = \dfrac{VA}{I}$ $V = \dfrac{42{,}000\,VA}{350\,A}$ **V = 120 V**
3φ	I	$I = \dfrac{VA}{V \times \sqrt{3}}$	72,000 VA, 208 V	I	$I = \dfrac{VA}{V \times \sqrt{3}}$ $I = \dfrac{72{,}000\,VA}{360}$ **I = 200 A**
3φ	VA	$VA = I \times V \times \sqrt{3}$	2 A, 240 V	VA	$VA = I \times V \times \sqrt{3}$ $VA = 2 \times 416$ **VA = 832 VA**

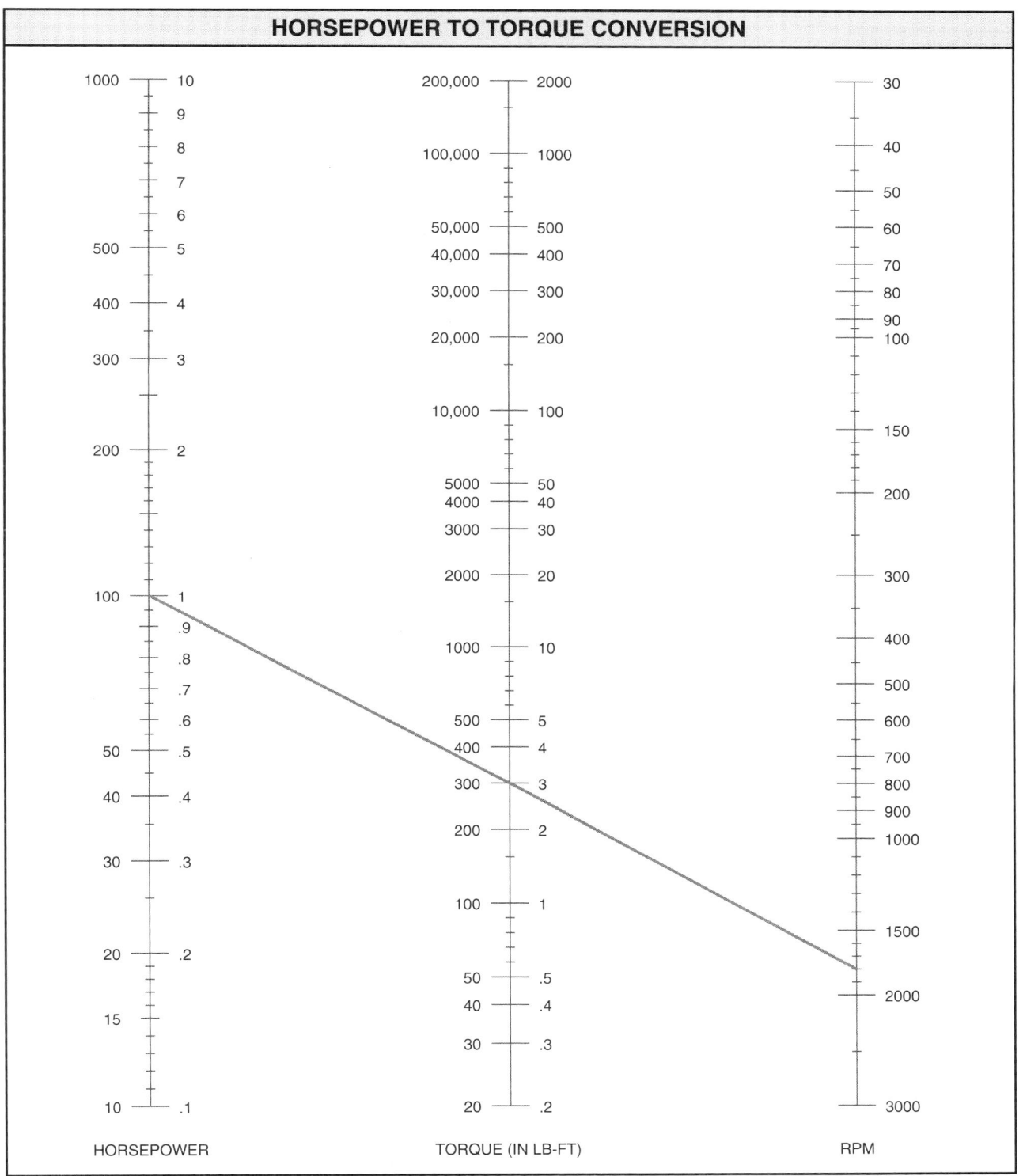

HOISTING EQUIPMENT CHECKLIST

1. Prior to Installation:
- ☐ Check for any possible damage during shipment. Do not install a damaged hoist.
- ☐ Check all lubricant levels.
- ☐ Check wire rope for damage if hoist wire-rope type. Be sure wire rope is properly seated in drum grooves and sheaves.
- ☐ Check chain for damage if hoist is chain type. Be sure chain properly enters sprockets and chain guiding points.
- ☐ Check to be sure that power supply shown on serial plate of hoist is the same as the power supply planned for connection to the hoist.

2. Installation:
- ☐ Install stationary mounting or trolley mounting to monorail beam exactly as instructed by the manufacturer's instructions.
- ☐ Check supporting structure, including monorail, to make sure it has a load rating equal to that of the hoist installed.

3. Power Supply:
- ☐ Make sure all electrical connections are made in accordance with manufacturer's wiring diagram, which is usually found inside the cover of the control enclosure.
- ☐ Make sure electrical supply system is in compliance with the National Electrical Code®.

4. Phase Connections:
- ☐ Depress the UP button on the pendant control to determine the direction of hook travel. If hook travel is upward, the hoist is properly phased. If it is downward, discontinue operation until phasing is corrected.
- ☐ Correct power connections if hoist is improperly phased by changing any two power line leads to the hoist. Never change internal wiring connections in the hoist or pendant control.
- ☐ Recheck operation of hoist after interchanging power line leads to confirm proper direction of motion.

5. Upper Limit Switch:
- ☐ Raise unloaded hook until it is approximately 1′ below the upper limit switch trip point. Slowly jog hook upward until hook can be raised no further. Lower block about 2′ and raise without jogging until limit switch trips and hook can be raised no further.
- ☐ Disconnect power supply and check all electrical connections if upper limit switch does not operate, or trip point is too close to hoist.
- ☐ Make any necessary adjustments.
- ☐ Reconnect power supply and recheck hoist operation after checking connections or making adjustments.

6. Lower Limit Switch:
- ☐ Check operation of hoist having a lower limit switch in same manner as for one with an upper limit switch. Never adjust lower limit switch to a point where less than one wrap of wire rope remains on the drum.

7. Lower Hook Travel (when hoist does not have lower limit switch):
- ☐ Lower the unloaded hook to its lowest possible operating point, or, for wire rope hoists, until two full wraps of wire remain on the drum.
- ☐ If it appears that less than two wraps of wire rope will be on the drum at the lowest possible operating point, the hoist cannot be installed or used unless it is equipped with a lower limit device.

8. Trolley Operation:
- ☐ Operate a trolley-mounted hoist over its entire travel distance on a monorail beam while the hoist is unloaded to check all clearances and verify that no interference occurs.

9. Braking System:
- ☐ Raise and lower hook, without load, stopping the motion at several points to test the operation of the brakes.
- ☐ Raise hook with capacity load several inches and stop to check that brake holds the load and that the load does not drift downward. If drift does not occur, raise and lower hook with capacity load, stopping the motion at several points to test the operation of the brakes.

10. Load Test:
- ☐ Load test the hoist with a load equal to 125% of the rated capacity load. If the hoist is equipped with a load limiting device that prevents the lifting of 125% of the rated load, testing should be accomplished with a load equal to 100% of the rated capacity load, followed by a test to check the function of the load limiting device.

11. Filing the Report:
- ☐ Prepare written report outlining installation procedures, problems encountered, and results of all checks and tests conducted. This report should indicate the approval or certification of the equipment for plant use, and should be signed by the responsible individual and filed in the equipment folder.

12. Operating Instructions:
- ☐ Issue instructions for hoist operators based on instructions and warning in hoist manufacturer's manual.
- ☐ Check warning tag or label on the hoist and make sure it stays there. Warning tag is a recent code requirement for new equipment. It is highly recommended for existing equipment. The warning tag should contain the following message:

WARNING:
To avoid injury, do not:
- lift more than rated load
- lift people or load over people
- operate with twisted, kinked, or damaged rope or chain
- operate damaged or malfunctioning hoist
- make side pulls that misalign rope or chain with hoist
- operate if rope is not seated in groove or chain in pockets
- operate unless travel devices limit function; test each shift
- operate hand-powered hoist except with hand power